Automechanic's Guide to Electronic Instrumentation and Microprocessors

Automechanic's Guide to Electronic Instrumentation and Microprocessors

Lynn Mosher

PRENTICE-HALL, INC., Englewood Cliffs, New Jersey 07632

Library of Congress Cataloging-in-Publication Data

Mosher, Lynn S.
   Automechanic's guide to electronic instrumentation and microprocessors.

   Includes index.
   1. Automobiles—Electronic equipment.
2. Microprocessors. I. Title.
TL272.5.M67 1987   629.2'549   86-30452
ISBN 0-13-054686-0

Editorial/production supervision and
  interior design: **Erica Orloff**
Cover design: **Lundgren Graphics, Ltd.**
Manufacturing buyer: **Rhett Conklin**

© 1987 by **Prentice-Hall, Inc.**
A Division of Simon & Schuster
Englewood Cliffs, New Jersey 07632

All rights reserved. No part of this book may be
reproduced, in any form or by any means,
without permission in writing from the publisher.

Printed in the United States of America

10 9 8 7 6 5 4 3 2 1

ISBN 0-13-054686-0 025

Prentice-Hall International (UK) Limited, *London*
Prentice-Hall of Australia Pty. Limited, *Sydney*
Prentice-Hall Canada, Inc., *Toronto*
Prentice-Hall Hispanoamericana, S.A., *Mexico*
Prentice-Hall of India Private Limited, *New Delhi*
Prentice-Hall of Japan, Inc., *Tokyo*
Prentice-Hall of Southeast Asia Pte. Ltd., *Singapore*
Editora Prentice-Hall do Brasil, Ltda., *Rio de Janeiro*

To Carol, for understanding and love.

Contents

	PREFACE	xi
one	**AUTOMOTIVE ELECTRONICS**	1

Introduction 1

What Electronics Is 1

Newer Electronic Technologies 3

Future Electronic Technology 5

two	**ELECTRICAL AND ELECTRONIC BASICS**	9

Automotive Component Operation 9

Electricity 12

Electronics 16

three	**HOW ELECTRONICS WORKS**	17

Electrical Conduction 17

Diodes 21

Transistors 24

Integrated Circuits 24

four	**PASSIVE OR LIGHT MODULATION DISPLAY SYSTEMS**	**30**

        Electronic Display Systems 30

        Liquid-Crystal Theory 31

        Liquid-Crystal-Display Construction 32

        Electrochromic Displays 37

        Electrophoretic Displays 37

five	**ACTIVE OR LIGHT-EMITTING DISPLAY SYSTEMS**	**39**

        Light-Emitting Displays 39

        Light-Emitting Diodes 39

        Matrix and Multiplex Circuitry 40

        Gas Discharge or Plasma Displays 43

        Vacuum Fluorescent Displays 44

        Cathode Ray Tubes 46

six	**MICROPROCESSOR TECHNOLOGY**	**50**

        Computer Basics 50

        Computer Component Terms 54

        Computer Operation: Registers and Addresses 55

        Computer Memory 56

        Automotive Computers: A Simple Program 57

seven	**MICROPROCESSOR-CONTROLLED SYSTEMS**	**60**

        Automotive Microprocessor Uses 60

        Engine Management 60

        Body Management 71

        Instrumentation Displays 72

eight	**AUTOMOTIVE SENSOR TECHNOLOGY**	**76**

        Engine System Sensors 76

        Automotive Sensors 88

        Sensor Buffers 94

nine **CONTROL TECHNOLOGY** **96**
    Function of Actuators 96

    Types of Actuators 96

ten **ELECTRONIC SERVICE** **102**
    Servicing Electronic Components 102

    Equipment Needed 102

    Service Philosophy 105

    Specific Service Examples 106

    GLOSSARY **119**

    INDEX **125**

Preface

The automotive industry has made monumental changes in the last 15 years. These changes include downsizing cars, integration into a world market, the initiation, development, and refinement of complex emission control, and the rapid development and expansion of electronics in many automobile systems. Each of these changes has a direct effect on the person servicing these new automobiles.

Automechanics of past decades, both the professional and the lay person, enjoyed the accomplishment of both understanding how the auto worked and being able to properly repair it. There was a genuine satisfaction in being able to understand how something worked. It was even more satisfying to be able to diagnose the cause of a failure and make the proper repairs. Many achieved improvements in their vehicle's operation over what the manufacturer had originally produced.

We have all found the automotive industry's rapid changes affecting our ability to continue productively as mechanics. If we are to continue to enjoy a feeling of accomplishment in the diagnosis and repair of these vehicles, it will require a dedication to learning the new technology.

This book is directed at a part of that problem; to bring together the basic theory and general practice of automotive electronics for those wanting to become current. It explores the basics in the electronic area for people who want to work on automobiles with confidence.

The first three chapters review where the automobile is today electronically, address the theory of basic electricity, and introduce the theory of electronics. Each of these is done specifically with the automobile in mind. Included also are many predictions of how electronics will affect our cars in the near and not-so-near future; more good reason for starting now to learn the technology.

Chapters 4 and 5 explain the theory and construction of the new and some yet to come display systems for automobiles. Step-by-step simple diagrams illustrate LCD, LED, matrix and multiplexing.

The basics of computers, microcomputers, minicomputers, integrated circuits and microprocessors, as well as the relationship between them and the many systems in autos controlled by a computer system of some type are all discussed in chapters 6 and 7. The purpose here is to have the reader begin to picture the relationship between various sensors, actuators, and a computer.

Chapters 8 and 9 describe the theory and physical construction of sensors and actuators being used or being considered for use in an automobile. Sensors are the part of the technology that makes the computer able to correct or improve the actions of any controlled system.

Chapter 10 first expresses the need for a new approach to diagnosis in the automotive field. Mechanical systems often lend themselves to the discovery of operational theory and diagnosis by disassembly. That is certainly not true with electronic systems. An understanding of theory, care and handling, and knowledge of the specific system are all needed to be successful. Most important is the right attitude; that of wanting to do everything correctly the first time. Electronic systems are not very forgiving. This chapter also gives some specific service examples.

Lynn Mosher

Automechanic's Guide to Electronic Instrumentation and Microprocessors

ONE

Automotive Electronics

INTRODUCTION

The automotive world is strongly into its greatest dramatic change; that change is the integration of electronics into all systems of the automobile. We are just at the beginning of the changes that will come because of electronics. "All systems of the automobile" means virtually that everything which once was a mechanical system will eventually be electronically enhanced or controlled.

We already have electric seats with electronic memories to return to a specific setting, mirrors that change position, coded door entry, resistance-sensing ignition key systems, electronic navigation, electronically controlled braking, controlled transmissions and suspensions, and that is just the start. Some automotive industry analysts predict that within just a few years, 30 to 40% of the cost of an automobile will be in the electronics.

WHAT ELECTRONICS IS

At this point, simply consider something to be *electrical* if most of the construction uses normal conductors such as copper wire with plastic insulation. For simplicity, *electronic* components can be considered those that use other than metallic materials for conducting electricity.

Figure 1-1 Electronic circuitry wired into composite circuitry.

A more complete discussion is developed in Chapter 2. Today, most electronic components use microcircuitry as the major unit, but with individual electrical and electronic components wired together to provide support (Fig. 1-1).

The types of electronic systems in common use in production automobiles over the past six to eight years include vacuum fluorescent displays, light-emitting diode displays, and more recently liquid-crystal displays. Vacuum fluorescent display clocks have that attractive green light usually seen in the segmented numerical system. In the last few years segmented letter displays have become more common in automobiles. Trip and miles per gallon display information has been packaged in the vacuum fluorescent mode (Figs. 1-2 and 1-3).

Figure 1-2 Vacuum fluorescent display. Courtesy B-O-C General Motors.

Speech synthesizing and more sophisticated sensing of temperature, pressure, light, and many other quantities are being introduced in an ever-increasing fashion.

The microprocessor is the "thinking" part of the system. Actually, the microprocessor simply but very quickly follows the program that is stored within it. Parts of the program "look" for inputs and allow the microprocessor to signal actuators to accomplish specific tasks if certain inputs are detected. Inputs can be sent to the microprocessor by switch input, sensor signals, or a keyboard input (Fig. 1-5).

Initially, these unique control systems were specifically designed into the system of which it was a part. The cruise control had its own microprocessor. The ignition system had its own microprocessor. Even the climate control system has had a separate microprocessor to maintain a specific temperature and humidity level.

The new trend is toward larger-capacity control systems that operate more than one subsystem. A body microprocessor system is now often responsible for heating and air conditioning, the instrument panel, the suspension system, and possibly, an integrated automatic brake system.

In addition to having larger computer systems, automobile manufacturers are beginning to connect the various computers by a data line that allows the information to be shared. This data line is being connected in a circular fashion so that even if the line is disconnected in one place, signals can continue in the other direction (Fig. 1-6).

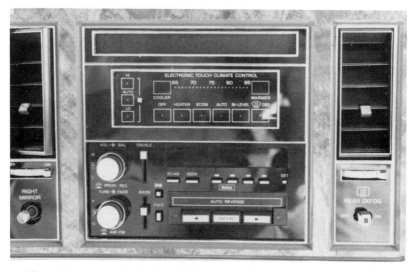

Figure 1-5 Automotive keyboard input system. Courtesy B-O-C General Motors.

Chap. 1 Automotive Electronics 3

Figure 1-3 Liquid-crystal display.

Other colors are available using light-emitting diodes; red, green, and yellow are all used. Liquid-crystal displays are now backlighted with almost any color desired. The theory and construction of each of these systems are discussed and illustrated in detail in Chapters 4 and 5.

NEWER ELECTRONIC TECHNOLOGIES

The use of small computers called microprocessors is now becoming common in all facets of automotive design and engineering. Clocks, trip information, speed control, ignition, fuel control, and many other systems make use of computer memory and direction. Cathode-ray-tube (CRT) display with touch-directed control is appearing (Fig. 1-4).

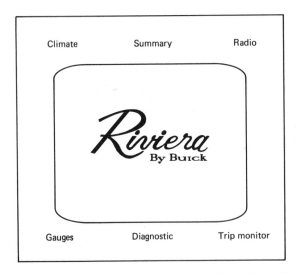

Figure 1-4 Cathode-ray-tube display. Courtesy of Buick Motor Division.

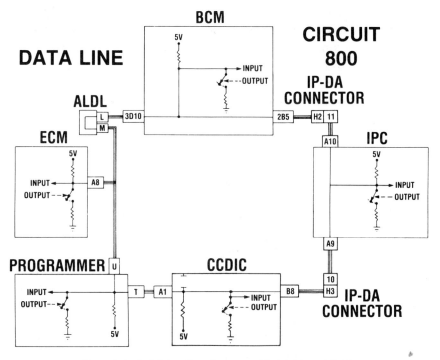

Figure 1-6 Serial data line. Courtesy of Buick Motor Division.

FUTURE ELECTRONIC TECHNOLOGY

Chassis

What will we see from the automotive industry in the coming years? "More than most of us are thinking about right now" is a conservative guess. Starting with the chassis, here are some things to look for. Many cars will go to four-wheel drive, using an electronic torque control to apportion power to each wheel. Wheel slippage will be controlled on acceleration as well as on braking. This will improve fuel economy and safety.

One of the big chassis changes will be four-wheel steering with electronic control. The steering actuators will be electric and the rear wheels will be able either to turn to follow the arch of the front wheels or turn parallel with the front wheels. Inputs to the computer programming for this steering system may include vehicle speed, lateral acceleration, vertical acceleration or road condition, and vehicle direction. Parallel parking could become a snap!

The electrical generating system may also be in for a few changes. You can expect to see a 24-volt charging system in the near future. It may come with just one battery or have a separate battery used only for the cranking motor. Cranking motors and most other actuators will make much greater use of high-strength permanent magnets, which are just beginning to appear in the automotive world.

Ride control will become far more sophisticated. Computer-controlled hydraulic and pneumatic suspension will displace metal and composite springs. Ride selection will include not only automatic shock absorber calibration, but also automatic vehicle ride height and possibly total control over body roll in turns. This could also mean an automatic increase in chassis clearance when going from on-road to off-road travel.

To reduce parasite air drag, the rear window of some automobiles will be eliminated. Rear vision will be accomplished by a television camera and a CRT screen. This will actually improve safety under many driving conditions. Rear vision at night will be improved, as the camera adjusts automatically for existing light conditions. You will no longer be blinded by the headlights of cars behind you.

Electronic sensing will warn you of cars in "blind" vision areas. When you need to make a lane change, the system will check to be sure that it is safe to do so. Road conditions will be monitored to determine if ice is present or how slippery rain has made the pavement. Forward sensing will possibly control automatic collision avoidance, putting on the brakes before you sense the need to do it.

Engine

Many of the systems that could be improved by electronic control already have been. But it goes without saying that those systems will continue to be improved. Better sensors, more efficient actuators, and more sophisticated programming are all on the way.

Mechanical control of engine functions that still exist will eventually be replaced by electronically controlled actuators. Engine valves will have solenoid control for greater breathing flexibility. That will eliminate the need for the camshaft. The oil pump may become a programmed electric pump that saves energy when less pressure is needed.

The cooling system is a candidate for early change. Coolant flow will be controlled by an electronically actuated electric valve. Very close control of engine temperature will improve fuel efficiency. Changes in engine loads will be responded to by the system much more precisely.

Direct fuel injection is another candidate for future electronic control. The higher fuel pressures required for injecting directly into the cylinder will require much shorter injection times or very small and precise injector nozzles. One of the advantages of this system will be much greater flexibility in the design of the intake system.

Direct-fire ignition systems will become common. Newer ignition systems will probably use one coil for each cylinder, right at the spark plug. The close proximity will reduce ignition problems. Ignition can occur on a cylinder-by-cylinder variation as prescribed by the computer program. This will allow for variances within the combustion of each cylinder.

Communication and Display

A look at the overall electronics system of the current automobile tells you that somehow the industry has to simplify the product so that service can be accomplished by the average mechanic. One step toward this is to get away from all the individual wires and connectors currently used in an automobile. Another problem to solve is the current practice of random location of various electronic components.

For a first approach to simplification, multiplexing will appear. Only three wires will be used to send power, to communicate, and to supply the ground for all components in the vehicle. Directions for a tail light, horn, or electric seat will be sent at different frequencies over the same common communication wire. Each light, actuator, or other device will have its own electronic receiver and allow power to be used off the central power line when the appropriate signal is received. Some receivers will only interpret the signal. Others will have intelligence and make decisions based on the information received. The computer can send hundreds of signals over the communication wire each second.

A step beyond multiplexing using conventional wire or shielded wire is to use fiber optics. Fiber-optic signals have almost no transmission interference, a wider bandwidth is possible, and they are not susceptible to corrosion. Fiber optics could be used to communicate directly from the driver input center, possibly with all controls mounted on the steering wheel, directly to all the systems requiring inputs.

One major advance that will take place is the central location of all computer systems. A main bus panel system will be located where it is easily serviceable. The panel will have standard slots or microprocessor connections. The connections will be on a "global" bus. All the parts that are plugged in will receive all the sensor and other signals sent. Each system will use whatever signals it needs to function. All information will be shared.

Each system that is ordered with a particular automobile will simply plug into the bus in any one of the slots and be able to communicate with the necessary sensors and actuators. Standardizing the mounting system is similar to the situation with many personal computers in home and office use today. Adding a printer or a modem is as simple as plugging it in. The bus system will give manufacturers real flexibility in the addition of options to any vehicle.

Diagnostics is another area open to considerable improvement. Future systems will be capable of greater depth in determining the actual cause of an "out-of-parameter" signal. In addition, the communication to the person receiving the information will be in the form of a verbal readout rather than by fault codes. Specific test procedures will be spelled out. This will require the mechanic to need far less dedicated test equipment, or no equipment. The system will advance toward total self-diagnostics. The diagnostic system will also be capable of testing sensors and actuators in a static mode. You will not have to run the engine or any other component to test them.

The display and communication to the driver are bound to get more sophisticated. Use of the CRT as an information center will become more standard. Complex navigation systems will proliferate. You will be able to locate all the roads, streets, businesses, parks, or any other location information you desire, right in the privacy of your own car. You might even expect to see commercials.

Greater use of liquid-crystal display (LCD) is coming. Flexible LCD display systems may be stretched across the entire width of the dashboard. More colored backlighting of display systems will be used. Color CRT systems are only a year or two away. Direct communication from satellites may be received while traveling.

The future of automotive electronics includes greater security for automobiles. The vehicles may unlock and start up only in responce to one of three or four specific voices. Voice commands may be used for many of the operations we now do by pushing switches or turning knobs. A final prediction is that the use of holographic images between the driver and the windshield will be used to convey driving information, such as speed and basic engine parameters, as well as any special messages that are required.

Well, that is at least some of the future that is not too far around the corner for the automobile. A lot of these systems will be here very soon. We must start to get ready for them now. Learning the basics is one way to do just that.

TWO

Electrical and Electronic Basics

AUTOMOTIVE COMPONENT OPERATION

It is easier to understand electrical and electronic component operation when either is compared to the operation of other systems. Some type of basic physical principle is involved in each operation. Let's start by briefly reviewing the basic energy conversion of a heat engine and some of the mechanical systems. A review of basic electrical theory will follow.

Heat Energy to Rotary Motion

Most automotive engines are of the four-stroke-cycle piston type. An air/fuel mixture is drawn into the cylinder, compressed, ignited, and burned to produce the power used to move the vehicle. The combustion process adds energy to the combustion products in the form of heat, which causes those molecules to move at greatly elevated speeds. The impact of the molecules on the surrounding surfaces causes the increased pressure that forces the piston down in the cylinder. The piston, being linked to the crankshaft by the piston pin and the connecting rod, causes the crankshaft to rotate with considerable force (Fig. 2-1).

The important principle is the addition of energy to the combustion gases from the chemical reaction to produce the pressure that moved the piston. The mechanical principle of leverage is also important.

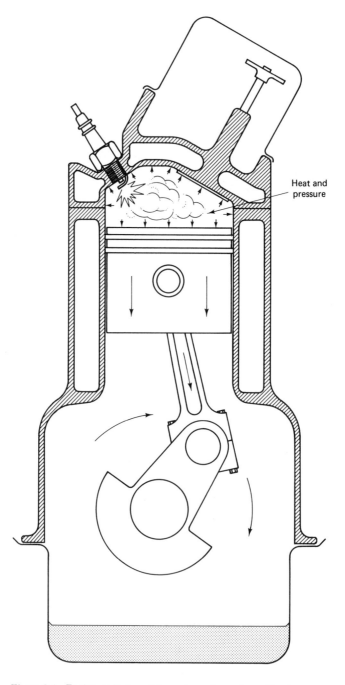

Figure 2-1 Engine, cylinder, piston, pin, rod, crank, and heat energy.

Chap. 2   Electrical and Electronic Basics

Movement is caused by combining an oxygen and a hydrocarbon molecule to form a new molecule and release the heat energy. This chemical reaction happens when the basic atoms of the existing molecules become rearranged to form the new molecules. Involvement of some of the basic material building blocks, the electrons, is primary to the production of energy and power (Fig. 2-2).

$$\underset{\text{Air}}{N + O_2 + H_2O + H_xC_x} \quad \underset{\text{Gasoline}}{} \longrightarrow \quad H_2O + CO_2 + NO_x + CO + H_xC_x$$

Figure 2-2   Chemical reaction of gasoline and air: products and heat.

Mechanical Energy Transfer

Stored energy in the form of chemicals, gasoline, and oxygen from the air, is converted by the automobile's engine into *power* in the form of rotary motion. The power is moved from place to place and used to produce movement of the automobile, or movement in its mechanical systems, or electricity in the alternator. Moving the energy from place to place is done with shafts, gears, pulleys and belts, hydraulic or pneumatic pumps, fluid, and motors or other actuators.

Power in the form of rotary motion can also be converted into electrical power by means of the alternator. Electromagnetic induction is the process involved. The rotating force is used to turn an electromagnet. The magnetic lines of force move with the rotating electromagnet, cutting through conductors (the copper wires in the stator). Electricity is produced in the conductor by the moving magnetic field passing through the conductor (Fig. 2-3).

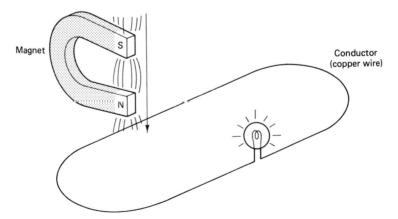

Figure 2-3   Magnet moving past a conductor with a light bulb.

ELECTRICITY

Basic Theory

Electricity is the movement of electrons in other than their proper or normal place. Electrons are normally situated in complicated orbits around the nucleus of the atom in a sort of cloud with layers. The various layers of the electrons designate at which energy levels the electrons are located, similar to clouds completely encircling the earth at various altitudes (Fig. 2-4). When the energy of the electrons in the outer cloud or ring of certain materials is increased, the electrons break loose and move to another atom. The electrons may be given that energy by moving a magnetic field past them, or they may be bumped out by electrons coming from another atom.

When enough electrons are moving in one direction to be a measurable force, we call that movement *electricity* (Fig. 2-5). Materials such as

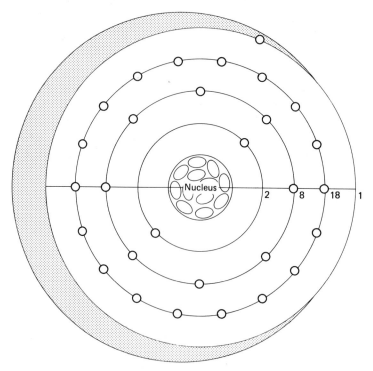

Copper atom cut into like an orange with one-quarter gone

Figure 2-4 Electrons in orbit: copper atom.

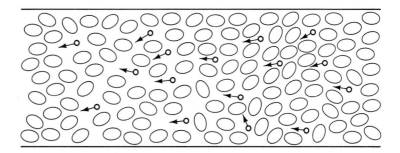

Figure 2-5 Electron movement in a material.

copper and aluminum have rather loosely held electrons in the outer orbits of their atoms. That fact makes these metals good electrical conductors.

Insulators are materials that have no loosely held outer electrons. All the electrons are held very tightly by their particular atomic structure. Good insulators include most plastics and ceramics.

Electrical Units

When electron movement occurs, names and values are attached to quantify that movement. The number of electrons going past a specific place in a given unit of time is valued in *ampheres*. Most technicians shorten that to "amps." An "amp" is a measure of the quantity of electrons that are in movement. The general movement of electrons in a specific direction along a conductor is also referred to as *current*. Current is given the symbol I. Amperes are given the symbol A. Amperes and current are the same quantity, so $I = A$.

The electrical force making the electrons move through the conductor is called *voltage*. Voltage can be compared to pressure in a pipeline. It is the push that makes the electrons move at the rate they travel. Technicians usually shorten this term to "volts." Another term used to describe this force is EMF or *electromotive force*, for which the symbol is E. V is used for volts. V and E are equal; that is, $V = E$.

Voltage can exist without electrons moving from atom to atom. The reasoning is similar to the idea of putting marbles in one end of a pipe. If there were a cap on the other end of the pipe and the pipe was already full of marbles, no more marbles could be put in the pipe. The force needed to put the marbles in may exist, but it is not having any effect. If the cap were taken off the end of the pipe, the force that

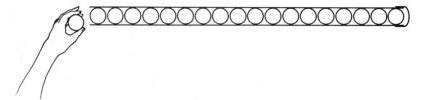

Figure 2-6 Trying to force marbles into a filled pipe with a cap on one end.

exists would determine how many marbles go into the pipe and how many come out the other end (Fig. 2-6).

In a 12-volt battery the chemical reaction that is ready to take place produces a force of about 2 volts per cell, adding up to a total push of 12 volts. When the electrons find a place to move to, the chemical reaction can continue and more electrons are broken free to become an electrical current.

A third quantity used to describe electricity in action is the opposition to electron movement through the conductor. This *resistance* to electron movement is given the quantitative value of *ohms*. The relative looseness of the electrons in the outer orbits is one determinant of the resistance of a given material. The slow or fast movement of the atoms themselves due to high or low material temperature is also a determinant of the resistance of a conductor. The cross-sectional size of the conductor is another factor that determines resistance. A larger volume of material for the electrons to travel through corresponds to a lower resistance. Resistance is given the symbol R. Ohms are given the symbol Ω. Here again R is equal to Ω.

The relationship of these basic electrical quantitative terms is given in *Ohm's law*. This law states that voltage divided by current equals resistance. Put in quantitative terms, $V/A = \Omega$. Another way to express the same law is $E/I = R$. Other algebraic expressions of the equation are $E = I \times R$ and $R = E/I$.

Another basic concept to understand when dealing with electricity is that of the *circuit*. The circuit is the path or paths that the electrons follow. Circuits are represented on paper using lines to represent conductors and various symbols to show resistances or other types of loads. The circuit shown in Fig. 2-7 is called a *series circuit* and includes the power source, a coil, two resistors, and a switch. The term "series" denotes that every component is in line with each of the others. The current traveling through one also travels through each of the others (Fig. 2-7). The switch must be closed for the circuit to be complete and allow electron movement.

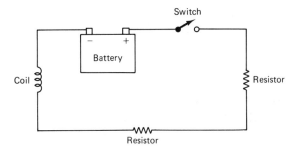

Figure 2-7 Simple series circuit with battery, two resistors, switch, and coil.

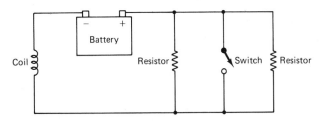

Figure 2-8 Simple parallel circuit with the same components but switch in parallel.

Figure 2-8 shows the same components in a *parallel circuit*. Notice that with the switch open or closed, current can still travel through the other components. Circuits may be as simple as those shown in Figs. 2-7 and 2-8 or may contain hundreds or thousands of components in parallel and series, as shown in Fig. 2-9.

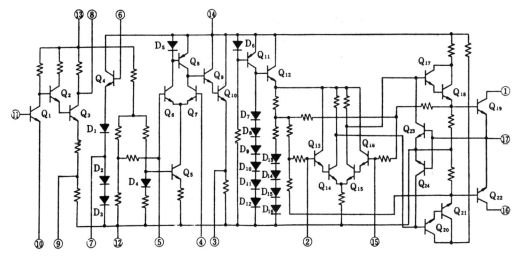

Figure 2-9 Complicated electronic circuit.

ELECTRONICS

Electricity deals with electron movement through conductors. Some conductors are "good", with a low resistance, whereas others have much higher resistances. An electric light bulb has a high electrical resistance, so the wire will be heated as the current travels through it, but it is still a solid metallic conductor.

The term *electronics* was first used when electrons were caused to move through a vacuum in a glass tube. The solid conductor no longer existed as the total electrical circuit. Since that time the term is generally used to describe electron movement in materials other than good metallic conductors and in some cases metallic resistors.

Semiconductors are materials that in their pure state would be very good insulators but have had tiny quantities of impurities added to create very narrowly defined electron conduction. Electronics may be said to include the study of electrical conduction in vacuums, certain gases, and semiconductor materials.

Today many of the popular electronic systems use silicone as the base material, "doped" to allow electrical conduction. The silicone may be formed in a very pure state by "growing" a large crystal under very controlled conditions. The crystal is then sliced into very thin sections. Many circuits can be manufactured out of each thin slice, which is divided into small "chips."

From this process we associate silicone chips with electronics. Most manufacturing designs and the assembly of devices such as stereo and television sets use electronic microcircuitry, individual electronic components (discrete), and electric circuitry connected together to produce the desired outcome. Electrical and electronic devices are used in combination, with electronics contributing the greatest number of components in the smallest space.

THREE

How Electronics Works

ELECTRICAL CONDUCTION

Good Conductors

In Chapter Two you were reminded that electricity is the movement of loose outer-orbit or shell electrons from one atom to another. When enough of these electrons are moving in one direction so that a current is measurable, we call that movement *electricity*. Electron movement occurs most easily in metallic materials. Most metallic substances are relatively good *conductors*. This means that they have the loose electrons in the outer shell, which allows easy electron displacement or movement.

 Displacement starts when a chemical or mechanical force pulls the first electrons out of their outer-shell positions in the conducting material. Nearby electrons in the conducting material are attracted to the positive charge of the unoccupied shell of those first atoms. The attraction creates a kind of domino effect of electrons jumping from atom to atom (Fig. 3-1).

Insulators

By contrast with conductors, *insulators* have complete or nearly complete outer electron shells. In that situation the electrons are much

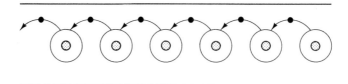

Figure 3-1 Electron movement in a conductor.

more strongly attracted to the nucleus and cannot be detached easily. When the electrons cannot easily be moved away from the atom, the electrons will not flow easily in that material. That material is an insulator. If a molecule is composed of two or more atoms, the relative looseness of the outer electrons still determines whether it is a conductor or an insulator.

Semiconductors

A third group of atoms, falling somewhere between good conductors and insulators, may be labeled *semiconductors*. If you look at a typical chemistry book, you will find that electron shells have a specific maximum number of electrons that can exist at each level. The first shell will hold two electrons, the second shell holds eight, and so on (Fig. 3-2). The number of electrons equals the number of protons in the nucleus.

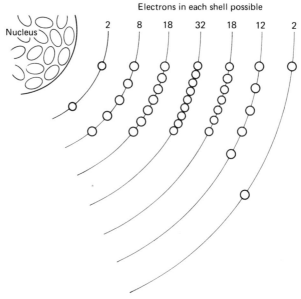

Figure 3-2 Electron shells: nucleus, 2, 8, 18, 32, 18, 12, 2.

Atoms with three or fewer electrons in the outer shell are conductors; atoms with five or more electrons in the outer shell are insulators. Atoms with four electrons in the outer shell are insulators that can become semiconductors. Silicone and germanium are the two primary elements used for semiconductors materials. These materials, when pure, act as insulators. However, when very tiny amounts of other atoms are combined with the four electron outer-shell atoms, electron movement becomes possible.

Figure 3-3 shows the form of pure silicone. You can see that the outer four electrons for each atom share orbits or shells with their neighboring atoms. This sharing of electrons creates a very strong bond and holds the electrons very tight. Electrical conduction will not occur easily.

Figure 3-4 has the same basic silicone structure with a small amount of indium included. The indium atom has three electrons in the outer shell. As the occasional indium atom occurs in the silicone structure, an

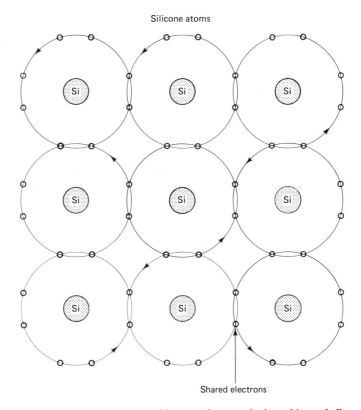

Figure 3-3 Silicone: atoms with outer electrons sharing orbits or shells.

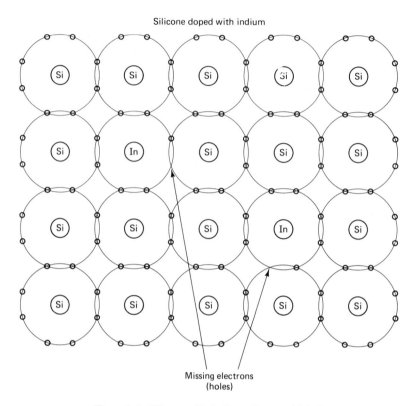

Figure 3-4 Silicone with indium: P-type with hole.

electron is *missing* in the shared bond. This missing electron makes it easy for a nearby electron to move into that place, leaving a hole for another electron to move into, and so on.

Conduction can occur as electron movement takes place. The ability of a semiconductor to conduct electricity depends initially upon the amount of "doping," or just how much foreign material is added to the pure silicone. The addition of arsenic to silicone creates a situation where there is an extra electron *added* to the silicone matrix for every arsenic atom added (Fig. 3-5). That electron is loosely held and allows the new material to become a conductor. Indium, arsenic, phosphorus, and boron are only a few of many elements used to dope semiconductor materials to gain various electronic effects.

You have already seen that the primary effect of doping is to create matrices having or lacking extra electrons. Those with extra electrons are called *N-type* semiconductors. *P-type* semiconductors lack electrons in the matrix. Conduction in P-type material is by "hole movement." As an electron moves into a hole, a new hole is created in the place the

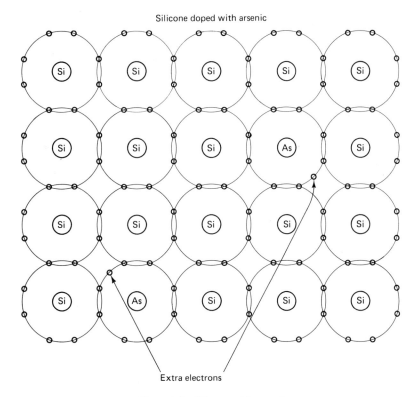

Figure 3-5 Silicone with arsenic.

electron just left. The holes move in the direction opposite to electron movement. Conduction in N-type material is by regular electron movement. Electrical conduction occurs easily in any direction in either of these materials by themselves. That is not the case when P-type material and N-type material are joined together.

DIODES

A simple solid-state electronic component is the *diode*. It consists of a piece of N-type semiconductor bonded to a piece of P-type semiconductor. Figure 3-6 shows what happens at the junction of the two materials.

The extra electrons that are loosely held in the N-type matrix are attracted to the spaces in the P-type matrix. Some electron movement occurs right at the junction, leaving holes in the N-type material. This creates a zone where positive and negative electrical charges are balanced.

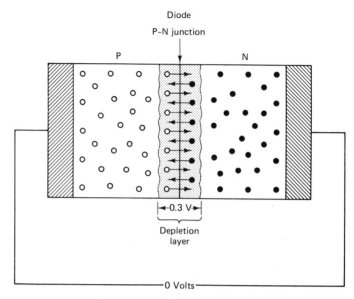

Figure 3-6 P-type and N-type materials at junction.

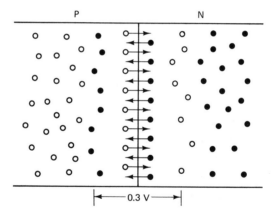

Figure 3-7 Depletion layer and diffusion potential across P-N junction in diode.

It is called the *depletion layer*. Just to either side of this zone are areas with excessive electrons (on the P side) or a lack of electrons (on the N side) (Fig. 3-7).

Diode Operation

Electrical conduction will not take place easily at the P-N junction. A voltage, called the *diffusion potential*, will exist across the very narrow depletion layer. No voltage differential exists, however, across the outer ends of the diode.

When a voltage is directed through the diode with the positive connected to the N-type end and the negative connected to the P-type end, the depletion layer is strengthened. No current will be conducted through the diode. This is called a negative or *reverse bias.*

However, if the voltage is connected positive to the P-type end and negative to the N-type end, the depletion layer will be eliminated and current will easily travel through the diode. The positive or *forward bias* causes the diode to conduct. Atoms on the N side are supplied with additional electrons from the power supply to fill the spaces left by the electrons as they move about, and they are now able to move through the P side of the diode.

Diode Rectification

The diode, because of the particular properties just described, acts as a gate that allows current to travel through it in one direction but not in the other. This special property gives the diode many important jobs,

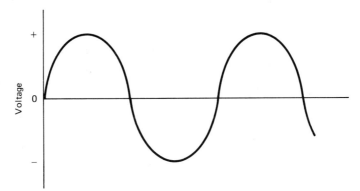

Figure 3-8 Sine wave with + 0 − over time.

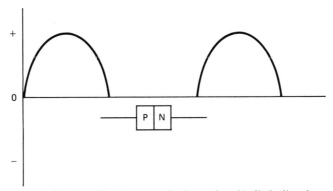

Figure 3-9 Rectified sine wave, both + and − with diode direction.

including alternating-current rectification. Alternating current and voltage consist of power that is first positive and then negative. The sine wave in Fig. 3-8 represents that situation. A single diode inserted in a circuit with an alternating current power supply will let electricity continue as pulsating direct current in the form shown in Fig. 3-9. Whether the rectified waveform is positive or negative depends on the direction in which the diode is connected into the circuit (Figs. 3-8 and 3-9).

TRANSISTORS

Transistor Operation

The *transistor* is a semiconductor device consisting of three bonded semiconductor regions. These regions may be arranged as PNP or NPN (Fig. 3-10). The center semiconductor part is called the *base*. It controls traffic through the transistor. The end that current goes into is the *emitter*. The end that current comes out of is called the *collector*. Conductors are connected to each section of the transistor (Fig. 3-11).

The transistor's general operation is shown in Fig. 3-12. With the switch to the base open, there is a depletion area at both N-P junctions. Current is available to travel through the transistor but cannot because of the depletion areas. Closing the switch to the base breaks down the depletion area to the left N-P junction and electrons shoot forward.

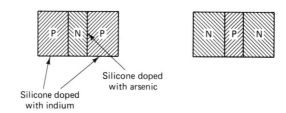

Figure 3-10 Transistor format: PNP and NPN.

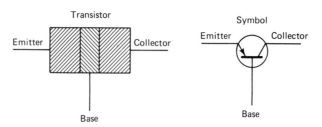

Figure 3-11 Transistor parts: labels and symbols.

Chap. 3 How Electronics Works   25

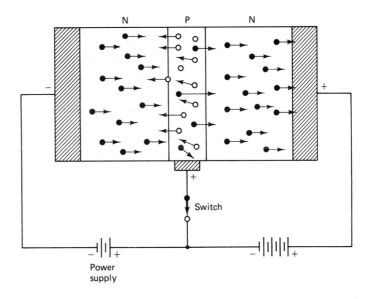

Figure 3-12 Transistor operation description, NPN.

Because of the thinness of the P semiconductor material, the velocity of the electrons, and the small size of the base connector, most of the electrons fly right into and through the second depletion area. Only a small proportion actually find their way to the base connection. If the base switch is opened, both depletion areas reestablish themselves, and current through the transistor is stopped.

In effect, any small current with the proper polarity that is conducted to the base will "turn the transistor on." The transistor allows a very small current to control a much larger one. Because the transistor is electronic and not mechanical in action, it can turn on and off thousands of times a second. This property gives the transistor a multitude of jobs to perform. It can be used, for example, as an electronic switch or an amplifier.

INTEGRATED CIRCUITS

Circuit Description

The purpose of an electrical or electronic circuit is to perform a specific function. A *circuit* consists of a group of electrical or electronic components connected together in a specific way. The circuit would then

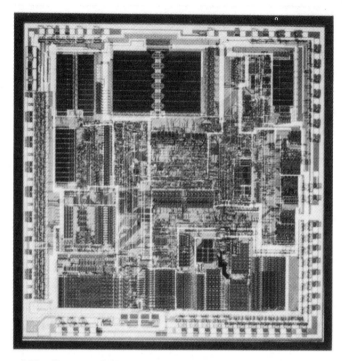

Figure 3-13 Close-up of IC chip. Courtesy "Solutions," a publication of Intel Corporation.

contain the components, their connecting conductors, and the specific pattern that would allow the function to occur.

An *integrated circuit* is a circuit that was manufactured within one continuous piece of semiconductor material. The components, their connections, and the circuit pattern were all designed in a way to make that possible. Typically, a circuit with hundreds or perhaps thousands of components can be designed and photographically miniaturized to fit in an area of ¼ square inch (Fig. 3-13).

Construction

The basic material in an integrated-circuit (IC) chip is very pure silicone. The surface of a thin flat piece of silicone is treated with many different processes to:

1. Change part of the surface to an oxide
2. Diffuse dopant atoms into the silicone
3. Deposit another semiconductor layer on the base material (epitaxy)
4. Deposit a metal layer on the semiconductor

Each of these material changes in or on the silicone base is added in a very precise pattern. This is accomplished using photographically reduced patterns called *masks*.

A simplified example of the process starts with applying a thin coat of special lacquer to the surface. Then the mask is set over the surface and ultraviolet light is used to expose the unmasked areas (Figs. 3-14 and 3-15).

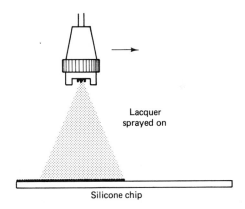

Figure 3-14 Lacquer on silicone base.

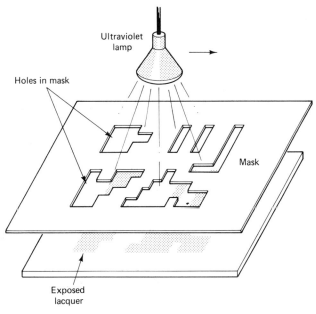

Figure 3-15 Mask applied and exposed to UV light.

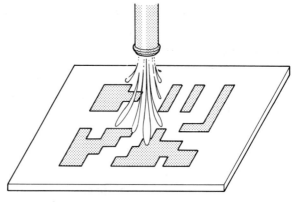

Silicone chip with exposed lacquer remaining, unexposed lacquer washed away

Figure 3-16 Solvent removal of unmasked lacquer.

A solvent then removes the unexposed lacquer coating from the remaining areas (Fig. 3-16). The uncovered silicone, silicone dioxide, or other surface is then etched to remove some of the semiconductor area. That newly exposed area may then be epitaxied, doped, metallized, or oxidized. This process may be repeated as many as 20 times to develop the complex circuitry required in layer after layer. Each layer may only be a few ten thousandths of an inch thick (Fig. 3-17).

The product of all this processing is a miniature complex circuit that is inexpensive, reliable, and able to be placed almost anywhere. Eventually, new automobiles may contain dozens and perhaps hundreds of the useful IC.

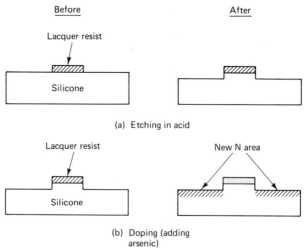

Figure 3-17 Processing of newly exposed surface: (a) etching; (b) doping; (c) metallizing; (d) epitaxy; (e) oxidizing.

Chap. 3 How Electronics Works 29

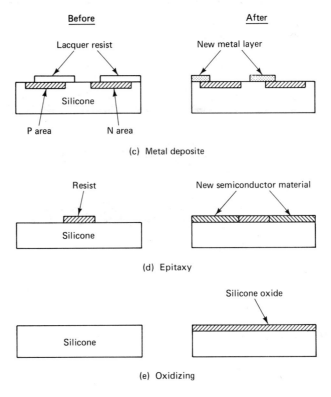

(c) Metal deposite

(d) Epitaxy

(e) Oxidizing

Figure 3-17 (continued)

FOUR

Passive or Light-Modulating Display Systems

ELECTRONIC DISPLAY SYSTEMS

Passive Displays

There are two classifications generally used to describe electronic display systems. The first type of display system does not produce light, but simply reflects light. This is a passive mode of conveying information. As you will see, this system works well when there is a high degree of ambient or natural light (Fig. 4-1).

Light-Modulating Displays

There are often several elements or layers in a passive display. Each element may have some effect on the quality or quantity of the light reflected. As the light is taken in and then reflected back out of the display, the light is changed or "modulated."

Figure 4-1 Close-up on a liquid-crystal display clock.

LIQUID-CRYSTAL THEORY

The term *crystal* or *crystalline* refers to a very orderly structure of molecules in a substance to form a geometrically shaped object. Common table salt and diamonds are examples of orderly molecular structure. Liquids, however, are generally understood to have a random molecular structure that moves to take the shape of the container.

One special group of liquids, primarily organic compounds, does exhibit qualities of molecular structure like those of a crystal. These liquids have an orderly molecular structure, thus the name *liquid crystal*. Included in this group are esters and biphenyls. Viewed from the side of a thin layer, the molecules in a regular liquid and in a liquid crystal might line up as shown in Fig. 4-2. You can see the orderly pattern of molecules in the crystalline liquid. The alignment of molecules in one particular liquid crystal, called *nematic*, is the one used for display systems. Other crystalline liquids, with different alignments, are not used.

When the molecules in a nematic crystalline liquid line up or exhibit their crystalline structure, light is able to pass through the liquid, even if the molecules themselves are opaque. If a mirrored surface is

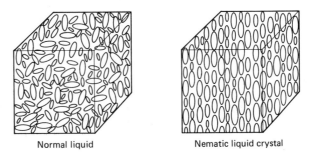

Figure 4-2 Molecular structure of a normal liquid and a crystalline liquid.

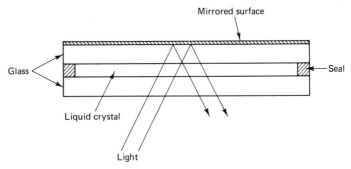

Figure 4-3 Light going through a liquid crystal and reflecting back.

placed behind the liquid crystal, the light will be reflected back through the liquid to the viewer (Fig. 4-3). If the molecules were not lined up, the liquid would be opaque and would absorb most of the light.

LIQUID-CRYSTAL-DISPLAY CONSTRUCTION

Simple Construction

The simplest form of liquid-crystal display consists of two thin layers of glass, each plated on the inner surface with a very thin layer of tin. The tin plating is so thin that it is *transparent*. Between the glass layers is a thin layer of crystalline liquid. The structure is sealed around all the edges with frit or polycarbonate (Fig. 4-4). Electrical conductors are connected to the thin tin plating on each of the glass layers. In the liquid's normal state, light passes through it. If the back side of the lower or back glass has a mirror plating, the light is reflected back out to the viewer.

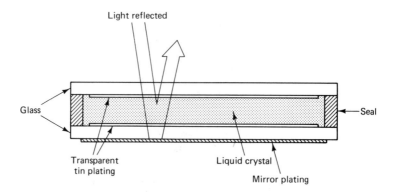

Figure 4-4 Liquid-crystal-display structure, glass–liquid crystal–glass with tin plating on each inner glass surface, light going through and reflecting off mirrored surface on back of back glass.

If, however, a voltage is applied across the two tin-plated surfaces, the orderly structure of the liquid molecules is disrupted and light will no longer pass through it (Fig. 4-5). The light is mostly absorbed by the liquid crystal. In this simple case the display would either reflect or absorb light across the complete surface. When the light is absorbed, the surface appears black.

Chap. 4   Passive or Light-Modulating Display Systems   33

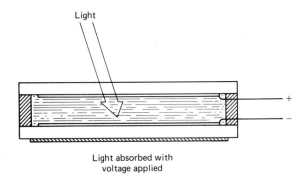

Figure 4-5 Voltage across liquid and structure, alignment disrupted, light absorbed.

Alpha/Numeric Displays

To make the display convey digital information, one of the plated glass surfaces must be changed. The tin plating on the inside of the back glass cover is put on in a seven-part (or greater) segmented pattern (Fig. 4-6). Each segment is connected by a pin or other electrical connector to the display drive system. The front glass plate electroplating is continuous across the display.

When an electrical potential is directed to one or more of the segmented plated parts, the molecular structure of the liquid crystal next to that segment is disturbed and light will no longer pass through it, as shown in Fig. 4-5. When the light is absorbed, the crystal liquid in front of that segment appears black. If all seven segments were turned on at once, the figure "8" would appear. If only segments 1, 2, and 3

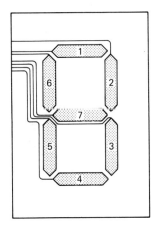

Figure 4-6 Seven-segment plating on inside back glass, segments numbered.

were turned on, the figure "7" would appear, and so on. A few of the alphabet characters can also be made with this construction. The letters A, H, L, U, and P are possible.

An alphabetical display requires more segments on the back glass plate. The 16-segment or "starburst" pattern will effectively do the complete job (Fig. 4-7). This does complicate the electronic system and electrical connections that drive the display. The even more complicated dot matrix and multiplex systems will be explained later.

Using these segmented systems produces a positive image. The figure appears where the light is absorbed. The background appears silver or gray where the light goes through the liquid crystal and then is reflected back through it again. Remember that the light is reflected off the mirror plating on the outside of the back glass cover. You can see why this type of display does so well in high ambient (natural) light. When little light is available, this display is hard to see. Also, because the viewing surface is really the back mirror or the liquid crystal and not the front surface, only about a 135° viewing angle is optimal (Fig. 4-8).

A negative image can be constructed with basically the same setup. The difference is that the electrical potential must be applied to the whole surface except to those information segments turned off to form the number. The negative image uses more power than the positive image. For that reason it is not often used for ambient light displays. One of the basic reasons for using the liquid-crystal display system in the first place is that this system has low power requirements. The

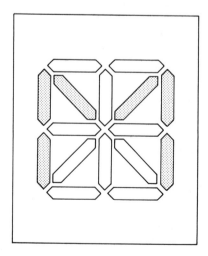

Figure 4-7 Sixteen-segment "starburst" alpha/numeric pattern.

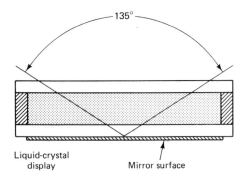

Figure 4-8 135° viewing angle.

power required for the liquid-crystal display is on the order of 2 to 3 V and around 0.005 A per segment. You can understand why the LCD is used so widely in watches and calculators.

Polarized Construction

Several added parts and changes are necessary to achieve optimal viewing with an LCD. The first change is in the liquid crystal itself. During the construction process of the cell, the liquid crystal is twisted 90°. That means that the molecules line up parallel to the glass covers. This is called a *twisted nematic* type of liquid crystal. Next, a thin layer called a *polarizer* is attached to the outer surface of the first glass plate (Fig. 4-9). The polarizer is a clear plastic part that lets light of only one polarity pass through it.

This outer polarizer cuts down on reflected light from the outer surface of the display. The polarizer, when aligned correctly, passes only the light that will easily go through the twisted nematic liquid

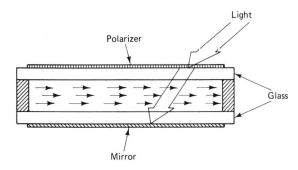

Figure 4-9 Twisted nematic LCD with polarizer on outer glass surface.

crystal. When this combination is used, the display will appear clear or transparent. The light is then reflected back out to the viewer.

When the voltage is applied to the conductive segments on the back plate of the display, the liquid molecules in that area change their alignment and the light is absorbed. That part of the display becomes black. It is possible to have the reflective part of the display (the mirror) reflect a particular color of light as well as silver or gray.

Backlighted Construction

Where ambient light conditions are not optimal, in automotive dashboard displays, for example, light must be supplied to the display. This can be done to make an LCD very attractive. The construction is similar to that of the twisted nematic display just discussed.

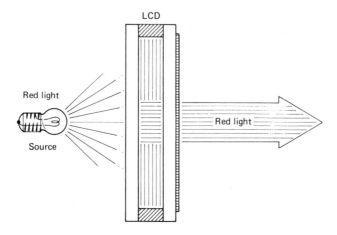

Figure 4-10 Backlighted negative-image LCD.

First, a light source is placed behind the LCD. No mirror coating is used on the back glass. The *negative-image* format is used in the display. Remember that in this case the background is kept opaque, either by applied voltage or orientation of the polarizer, and the display or number segments are allowed to let the light through. When a particular number segment has voltage applied, the liquid molecules line up in a way that lets the light come through from the back to the front (Fig. 4-10). Any color of light may be used as the backlight source. The display has a positive lighted image. This is also called a *transmissive display*.

Dichrolic LCD

Another LCD type may be seen in some systems. The *dichrolic* format involves using a dye together with the liquid-crystal material. The dichrolic dye is mixed with the liquid crystal but does not change its molecular structure. When ambient light is applied to the surface of the display, certain wavelengths are absorbed. What you see is a colored background. The color, depending on the dichrolic dye used, can be orange, yellow, or many other colors. When voltage is applied to the positive-image LCD, the numbers appear black on the colored background.

ELECTROCHROMIC DISPLAYS

One other possible display system would use the characteristic of a material that changes color when voltage is applied. This change is called *electrochromism*. One of the features of iridium as an electrochromic material is that it changes from clear to blue-black when voltage is applied but does not change back when the voltage is removed. A reverse voltage must be applied to clear the display.

This material is not in a liquid form but would be placed in a thin solid layer between two glass plates. The front plate might have a complete layer of transparent conductive coating, while the back plate would receive the segmented conductive coating to form the digits. Connectors from the conductive coating segments and the front conductor to the display driver would provide the path for the needed voltage.

A brief pulse of voltage between any segment and the front plate would turn the color on, and a pulse of reversed polarity would return that segment to the clear format. If the reverse polarity voltage were not applied, the display would stay "on" but no electrical energy would be used.

ELECTROPHORETIC DISPLAYS

The theory of movement of charged particles in a liquid by applying a voltage is described by the term *electrophoresis*; the term EPID is also used. This display system does not look very different in construction from the LCD unit. There are two glass plates, both coated with transparent conductor coatings. Again the rear coating is in the familiar pattern of seven or more segments used to form numbers or letters.

Between the glass plates is placed a solution of one color and very tiny particles are mixed in, with a negative charge. These tiny particles have another contrasting color. When voltage is applied across the segmented and front conductors, the charged particles are driven to the front plate, but only in front of the driver segments. What you will see is the solution color background and the contrasting colored particles shaped in the form of the activated segments.

When the polarity of the electrical energy is changed, the contrasting colored charged particles will then go to the opposite electrical surface. The colored liquid will now be seen at the front glass. The display is possible in a number of different-colored fluids and contrasting charged particles. Because the light is reflected off the front glass plate, this display can be seen from very wide viewing angles (Fig. 4-11).

These passive displays, with the exception of the backlighted LCD, do not produce any light. They simply modify and/or reflect the light directed at them. External lighting would have to be supplied to view them in a darkened area or at night.

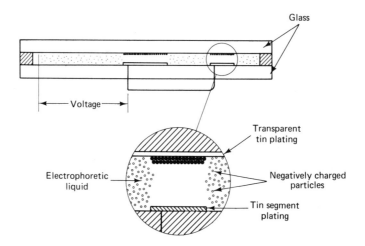

Figure 4-11 Electrophoretic display theory.

FIVE

Active or Light-Emitting Display Systems

LIGHT-EMITTING DISPLAYS

In these systems light is emitted from each active element of the display. It should be obvious that an active system will require more power to operate. It takes power to produce light. For many systems there is also the possibility of more complicated drive components. Either higher voltages or currents or both may be required to produce the emitted light.

Another potential complication is that the driving-circuit IC chip may not be able to handle the higher voltage or current. Additional amplifiers might be necessary to activate each display element. Here is a look at the operation and construction of active light-emitting display systems.

LIGHT-EMITTING DIODES

The name "diode" tells you that this is a solid-state device. In this case the display element consists of a P-N-junction diode. It is made of a material such as gallium phosphide. When forward bias (voltage) is applied, the diode emits light (Fig. 5-1). A voltage as low as 3 or 4 V may be enough in some designs to produce the desired luminescence. Brightness can be increased by increasing voltage, to a point. Different

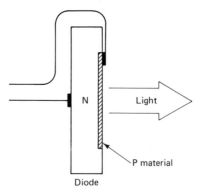

Figure 5-1 Light-emitting-diode construction.

combinations of P and N materials will produce different colors of light. The red-light-producing diodes are said to be the most efficient electrically.

Some advantages of LEDs are that they are very long lived, reliable, operate well over a wide temperature range, and respond quickly. One way to produce a display panel using LEDs would be to arrange them in either the seven-segment or starburst patterns shown in Chapter 4. That system would require from 14 to 48 different electrical connections for each display letter or number. A 10-letter display might require close to 500 connections. One important advantage of LEDs is their ability to be used in a matrix and a multiplex circuit.

MATRIX AND MULTIPLEX CIRCUITRY

Using any display device in a matrix circuit depends on the device's *threshold*. The threshold is the energy it takes to activate the display element. The signal used to activate the display element is sent along conductors. In the case of a light-emitting diode, a different bias must be signaled to the P-N junctions.

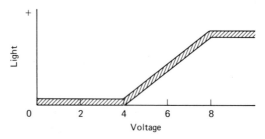

Figure 5-2 Graph showing voltage applied to light emitted.

We would say that the difference between the positive and negative leads is the voltage signaled. When this voltage is slowly raised, a point will be reached where the LED begins to emit light. That voltage is the threshold voltage for that diode. As long as the voltage signaled to that diode is below that threshold, no light will be emitted (Fig. 5-2).

Matrix

Now let's construct a simple matrix with a 5 by 7 grid. This would contain 35 LEDs connected together as shown in Fig. 5-3. The seven horizontal leads might all be connected to the N sides of each diode. The five vertical leads would then all connect to the P sides of each diode. In the most simple case a signal is sent through leads 1N, 5N, 2P, and 4P. The signal from any one lead is less than threshold for each diode.

However, where the signal is applied to any diode from two different sources, both N and P, the added signal strength is above the thresh-

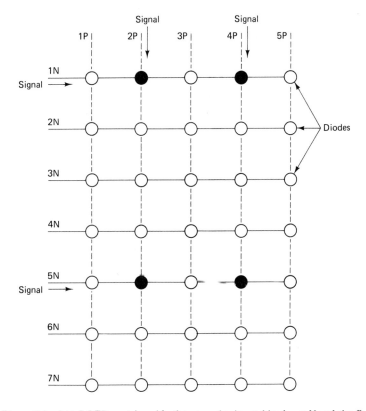

Figure 5-3 5 × 7 LED matrix grid; the seven horizontal leads are N and the five vertical are P.

old and that diode emits light. That is how the matrix part of the puzzle operates.

Multiplex

Multiplex means to send several signals over the same conductor or channel at the same time. For this display system example, multiplexing will be sending many signals in rapid succession over the same conductors. Several new components are necessary for this to work. These will be shown in block diagrams, as it is only necessary to understand what they do, not how they work.

Let us assume that some other system has determined what character is to be displayed. For example, the climate control computer has taken an analog voltage from a temperature sensor and is signaling the display to show the updated value (Fig. 5-4).

The display character selector receives a digital signal that is interpreted as a specific character. The character selector then signals the timekeeper to start sending to the driver the various component parts of the letter to be formed. These signals come at very precisely timed intervals. The display system driver then sends the appropriate voltage along the P and N leads of the display.

Each signal is sent for only the tiniest fraction of a second. Let's say that the signal is sent first to 1N and 3P; then it is sent to 2N and 3P; and so on. Each diode lights up in turn. When the last diode that makes up the character to be displayed is lit, the sequence is repeated all over again. We see the diodes being sequenced *as if they were all lit together*! The sequence is repeated so fast that we do not see even a flicker. The diodes look as if they *are* all on together.

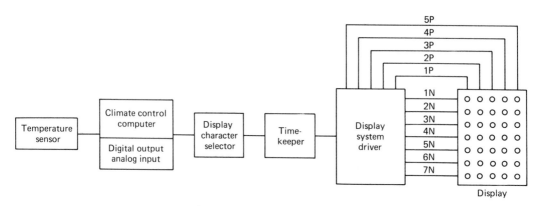

Figure 5-4 Block diagram of display system.

The advantage of multiplexing is obvious. Only 12 wires are needed to light 35 diodes. Manufacturing of that part of the display is much simpler. Also, when the display is being used, there is a considerable energy savings. The lower current use of the multiplexed system will make batteries last longer. Multiplexing can be accomplished with several seven-segment numeric-type displays at one time also. Matrixing and multiplexing make a very useful display system possible at a reasonable cost.

GAS DISCHARGE OR PLASMA DISPLAYS

A plasma occurs when a voltage is applied to a gas, causing ionization of the gas. Ionizing is the removal of an electron or electrons from the outer shell of the atom, causing the atom to have a positive charge. Some gases, neon, for example, glow when ionized. The voltage applied to the neon atom may actually break it, releasing two electrons. These electrons smash into other neon atoms, exciting them to higher energy levels. When their return to the regular energy level is made, the extra energy is released in the form of light. When the gas glows, it is called a *plasma emission* or *display*.

The construction of a display device consists of a glass container filled with the special gas. An anode and a cathode are fitted into the glass tube so that an electrical charge may be placed through the gas (Fig. 5-5). Some of the most familar examples of this technology are the "neon" lights used throughout the world for advertising purposes. The store displays of this type that you see use a high-voltage ac power supply. A transformer is used to change 120 V ac to about 6000 V ac.

The smaller gas plasma display systems considered for use in automotive applications require a minimum of about 170 to 180 V. The

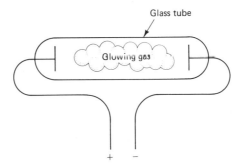

Figure 5-5 Simple plasma display tube: anode, cathode, and gas.

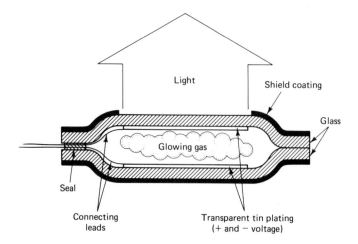

Figure 5-6 Automobile-type gas plasma display tube: transparent anode and segmented cathode with pin connections.

construction still consists of a glass tube, but the tube is flattened and has a transparent anode coating on the front face. The cathodes have the now-familiar seven or more segments to form the number or letter required of the display system (Fig. 5-6). This system is obviously different from the LED display, as there is a glass tube filled with gas. Also, it is not solid state. But the PGD (planer gas discharge) display is not like the incandescent bulb either! There isn't a fragile filament to get broken at the smallest bump or bang.

Some advantages of PGDs are their ability to be multiplexed, their bright appearance, and the availability of orange and green colors. Disadvantages include the high voltage required and a more complicated drive system than the LCD or LED.

VACUUM FLUORESCENT DISPLAYS

There are three required elements in the basic vacuum fluorescent (VF) display. There is a filament-type cathode, a grid, and a phosphor-coated anode (Fig. 5-7). When a voltage is applied to the filament cathode, it heats up, releasing electrons. These electrons are accelerated by and through the control grid to the phosphor-coated anode. The electrons striking the phosphor cause the coating to emit light.

The voltage applied to the printed phosphor elements of the anode can be used to determine which anode element will receive the accelerated electrons, and therefore which will emit light. The anode elements are arranged in the familiar seven-segment (or greater) configuration,

Chap. 5   Active or Light-Emitting Display Systems

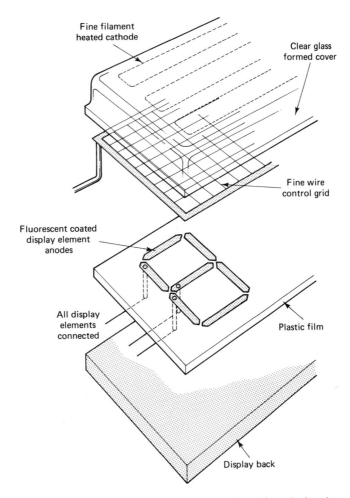

Figure 5-7 Basic VF elements: filament cathode, grid, and phosphor-coated anode on glass.

with the alpha/numeric character on the inside of the viewing surface of the display. This assembly is contained in a glass envelope that is evacuated. As the phosphor coating that emits light is on the inside of the foremost surface, the display can be seen from a fairly wide viewing angle.

Two additional advantages of this type of display account for its wide acceptance. First, the anode can be developed in the matrix format. With an x/y grid it is possible to multiplex the display and provide a variety of information with the minimum number of leads.

Second, the color of light emitted is a very pleasant green which is right in the middle range of the eye's ability to discern light. The eye perceives the light as brighter than it actually is compared to other colors. Less power is required to make the display comfortably visible compared to other colors.

Currently, the VF display system is the most used system in automobiles. Voltage requirements in some uses are as low as 10 V. Vacuum fluorescent technology will be found in auto clocks, speedometer segments, and most general instrument displays.

CATHODE RAY TUBES

We have all watched information presented on the cathode ray tubes that are used in television. The information conveyed is pictorial, written, can be in color, and can change continuously. The cathode ray tube (CRT) has the great flexibility needed in an information display system. That system is coming to automobiles. Let's look first at the basic operation of the CRT.

The outer envelope is an evacuated lengthened glass tube. The display end of the tube is broad and flattened and the other end is necked down to a narrow tube. The assembly at the back or narrow end of the tube consists of an electrical heating element, a cathode, and one or more accelerating anodes (Fig. 5-8).

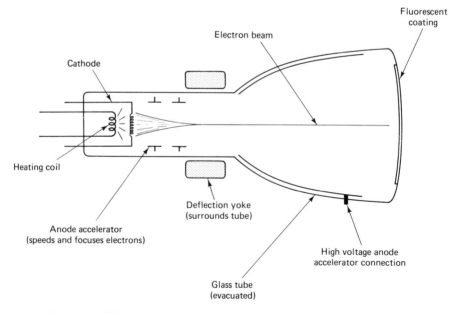

Figure 5-8 CRT construction: tube, phosphor coating, heating element, cathode element, anodes, magnetic coils (yokes).

Just before the tube begins to open out, there are one or more electrical coil(s) surrounding it. The inner front surface of the tube is coated with one or more types of phosphor.

CRT Operation

The basic operation of the CRT is as follows. Electrons are set free as a voltage is applied at the cathode. The heating element behind the cathode is an aid to this function. The electrons are accelerated out toward the phosphor-coated screen by the voltage applied to the anodes. The electrical coils surrounding the tube produce a magnetic field that concentrates or focuses and directs the stream of electrons toward a specific point on the phosphor coating. When the electrons strike the coating, it emits the light seen at the front face of the tube.

The deflecting coils or "yokes" are controlled by the voltage applied to them. Varying the energy to two different yokes changes the impact point of the electron beam. Precise control of this function can create a light-emitting letter, number, or picture on the phosphor-coated surface of the "screen."

Analog CRT Control

Two additional factors must be explained to understand the basic operation of the CRT display system. The first is analog control. The image you are used to seeing on your TV is produced by scanning. That means that the electron beam is swept across the screen in a precise pattern that is repeated at a consistent frequency (Fig. 5-9). The frequency is

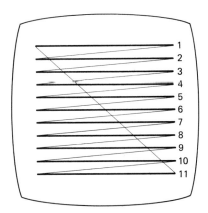

Figure 5-9 Scan pattern of TV.

fast enough that the brightness of a given phosphor spot will not decay between the time when the electron beam leaves and returns.

Second, at the same time, the control system for the cathode emitting the electrons is regulated to increase or decrease the beam intensity in each spot to produce the information needed to be displayed. More electrons emitted produce a brighter spot, and fewer electrons produce a darker spot on the screen. Control of these two systems together produces the image that we see.

Digital control of the electron beam is also possible for a CRT display system. In this system a scanning format is not used. The beam is directed to specific areas of the screen in the somewhat random pattern needed to produce the specific image. This produces a brighter image but requires a more complicated control system. As you might guess, the specific control systems are far more complicated than it is our intention to cover in this book. People who wish to go more deeply into CRT systems will have little trouble finding materials.

Additional advantages of the CRT display system include the ability to communicate very complex information, the ability to use multiple colors in the same display, touch-screen feedback capability, and multiple outputs in different places on the screen at the same time. Two disadvantages are the higher cost and the fairly large space requirement of the display.

CRT Uses

Current uses of CRT in automotive display systems include the navigation concept, the overall information display, and special information displays. The navigation concept may include a receiver that gathers transmissions from a local transmitter or a satellite transmitter. Magnetic discs or tape also can provide information. The information displayed would include a map of the area surrounding the auto's current position and where the automobile was at that exact moment. Larger area maps could be brought to the screen, as well as computed information, such as time of travel to specific locations or locations of gas stations or other establishments.

In the overall information display the CRT takes the place of all the other instrument displays. Speed, engine temperature, oil pressure, odometer, and all information currently found in the dash display would appear in different locations on the CRT. Information that was out of parameter, such as high engine temperature, might be brought to a more prominent location on the CRT. Voice alert will probably also be used in conjunction with the special modes. Additionally, touch control of the screen can be used to bring up information or to control automotive systems, such as the air conditioning or lights.

Special information displays could include entertainment data, vidio transmission from the rear as a substitute for rear-view mirrors, and auxiliary equipment information. Systems such as the seat belts in all seats, climate control, weather information, and road conditions could be displayed on special information CRT displays.

SIX

Microprocessor Technology

COMPUTER BASICS

The computer is made up of a group of interconnected components. One way to simplify the overall concept is to illustrate it in a block diagram. Each block represents one component of the computer. The interconnecting lines represent the way each component is connected to others. Communication from component to component can occur only over the existing connections (Fig. 6-1). The arrows on the lines represent the directions in which communication can take place.

Clock

The clock in a computer sets the pace at which processing occurs. It is the basic input to the control unit.

Control Unit

The control unit (CU) directs the other parts of the computer. Memory contains a specific set of instructions that the control unit follows. Each instruction is followed in sequence as the control unit directs each of the other computer components to perform one specific function at a time. An example of control unit function is as follows:

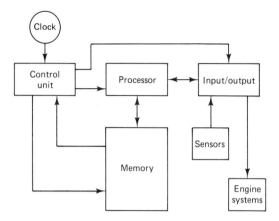

Figure 6-1   Block diagram of a basic computer.

1. Look at memory address A1.
2. Then direct the processor, input/output, or memory to perform the function stored in A1.
3. Accept clock input.
4. Look at memory address A2 . . . (through all the memory addresses in that sequence).
68. Return to memory address A1 (sequence starts over).

You can see from the example that the CU is the traffic cop of the computer. Its job is to take directions from the instruction set in that part of memory that is specifically set up for the controller. From those instructions it directs the traffic in a specific sequence through the computer components (Fig. 6-2).

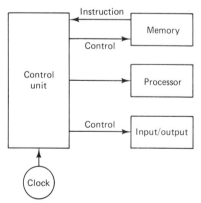

Figure 6-2   Block diagram of the control unit with connections.

Processor

The processor is generally the place where the actual functions reside. It is where numbers are crunched or decisions are made. The design of the processor determines whether it can add or subtract, determine if a number is larger or smaller than another, or branch into another direction. These functions are usually given names, such as test/branch, input/output, logic, arithmetic, and load/store (Fig. 6-3).

If addition or any other math function is needed, and it is designed into the processor, that is where that step is accomplished. When the control unit tells the processor to store incoming information, the processor finds specific addresses in memory for the incoming data and places it in those addresses.

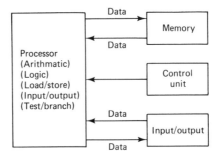

Figure 6-3 Block diagram of processor with connections.

When a signal from the input/output unit comes to the processor, the processor may be required to determine if the data are outside the stored parameters. If so, the processor may cause a branch function to occur. The processor will compare the input number to a specific number held in memory. If the number is higher (or smaller, depending on the program), it will then branch to a different routine (also depending on the program in the memory). That process is called *test/branch*. Each of these functions is performed by the processor as directed by the control unit.

The processor both accepts data from the input/output unit and sends data back out to either the memory or the input/output unit. Data are sent to the memory to specific addresses for temporary or permanent storage. Data are sent to the input/output unit, where they are converted and sent to specific actuators or sensors.

Input/Output

This unit has the responsibility of accepting data from sensors or other types of inputs and converting those data to an acceptable form to send on to the processor. The input/output also accepts data from the processor and converts them into an acceptable form to be sent out to the actuators. The input/output unit (I/O) also takes orders from the control unit concerning when to send outputs to the actuators or when to accept inputs from the sensors (Fig. 6-4).

Input into the I/O from the sensors is usually of the analog type. That means that the sensor signal is possibly either a variable voltage or a variable resistance. The variable resistance actually varies the voltage or current sent out by the I/O initially. The signal might also be a set

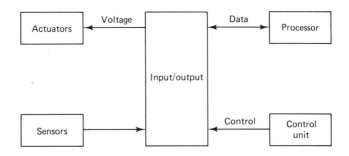

Figure 6-4 Block diagram of input/output and connections.

voltage either on or off. The input is not, however, in a digital format readable by the processor. The I/O's job, or that of the "buffer" before the I/O, is to make that conversion (more about buffers later). Each specific variable that is accepted by the I/O is converted to a specific digital signal. A specific analog input will always be given exactly the same digital output: +0.23 V from the oxygen sensor is converted to the same specific output to the processor every time.

The second job of the I/O is to signal the actuators to function properly. Given data from the processor on the length of time to remain open and from the control unit when to open, the I/O sends the proper voltage to the fuel injector to open it and hold it open for the correct length of time. Given data from the processor and control from the CU, the I/O sends the proper voltage to the coil to provide a spark to the correct cylinder at the exact time for optimal engine operation.

Each component of the computer is directly or indirectly tied to the others. This relationship is necessary for any of them to function. The communication flow must follow the designer's directions for the computer to function properly.

COMPUTER COMPONENT TERMS

Integrated Circuits

Computers are constructed primarily with integrated circuits. An integrated circuit is the combining of many thousand individual components, such as resistors, capacitors, transistors, and diodes, into one small piece of semiconductor material. This process is accomplished by starting with very pure silicone in the form of a flat thin disk.

Prints of the final designs chosen for the various components to be constructed are photographically reduced to tiny but very precise images. These images are imposed on a chemical resist placed on the surface of the silicone. Processing removes the silicone, adds more

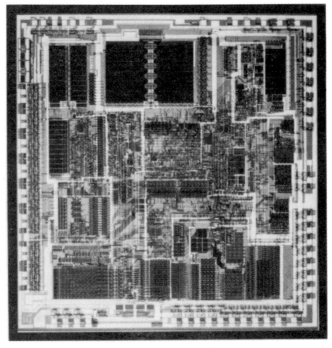

Figure 6-5 Close-up of integrated circuit.

crystalline structure to the silicone, and dopes the exposed silicone or deposits metallic conductor material on the silicone.

These processes, each performed in a specific sequence, form the various interconnected components that make up the integrated circuit. This is all accomplished on a miniature scale using photographic reduction to produce the designs needed (Fig. 6-5). Many thousand components may be constructed on a single 1/4 square inch silicone chip by this process.

The design of integrated circuits has become so sophisticated that one chip may perform only one function, or it may in fact be a complete computer. The single-chip microcomputer is usually fairly limited in its capacity.

One design would be to manufacture one integrated circuit as the control unit. Another chip would be designed for the processor, and yet another for the major part of the I/O unit. It is common, however, to manufacture the control unit and the processor as one chip. This chip is called a *microporcessor*. A computer might use one or several IC memory chips for its memory.

Microcomputer

A microprocessor with added memory and I/O circuits becomes a *microcomputer*. These are usually connected together on a circuit board to be used within a specific manufactured product.

The microcomputer is now ready for custom design for use in an automobile. A program could be manufactured into the memory. Other ways of installing the program in memory will be discussed shortly.

Minicomputer

The *minicomputer* is the "complete" package. It has a proper power supply, software, and a way to communicate with its user: through a CRT, a keyboard, and a printer. The minicomputer does, of course, contain a microcomputer.

COMPUTER OPERATION: REGISTERS AND ADDRESSES

The memory of a computer must be arranged in a very logical order for the unit to function. All the information stored by the memory is first

A	1010	1	0001
B	1011	2	0010
C	1100	3	0011
D	1101	4	0100
E	1110	5	0110

Figure 6-6 Binary code for 0–5 and A–F.

put in the form of binary numbers, 0 and 1. Each number or letter or other symbol has a binary code. Figure 6-6 illustrates part of that code.

Using only two values, 0 and 1, the computer can make the decision to put a circuit either "off" or "on." A switch that is "off" is interpreted as 0 and a circuit that is "on" is 1. The binary digit, 0 or 1, is called a *bit*. If the computer is capable of processing 8 bits at a time, it is classified in terms of its *byte* capacity. An example of a byte is the code 11010010, which is the binary code for D2. A computer's memory is a large collection of *electronic switches* that store the binary code signals. Newer computers can process 16 or 32 bits at a time.

Each storage place is called a *register*; thus memory is a collection of registers. Each register is given a numerical *address*. As the program is followed, the control unit or the processor will fetch the stored information from each numbered address. There are different types of memory that allow a computer to perform its designed function.

COMPUTER MEMORY

Primary Memory

Primary memory consists of two types of memory, read-only memory and read/write memory. *Read-only memory* (ROM) usually holds the permanent directions or instructions and data by which the individual computer is governed. *Read* means to find out what number is stored in a specific address. *Write* means to put a new number into a specific address. As the name implies, the microprocessor of a computer cannot change the data stored in its ROM. That is not to say that the data in a ROM can never be changed. The memory that remains after power is turned off is "nonvolatile" and is referred to as *firmware*.

The *read/write* (R/W) *memory* is designed to hold changing data. To change these data, very specific directions are required in the program. These are still data that the microprocessor uses to function correctly.

The ROM binary numbers can be placed in as part of the manufacturing process. The IC manufacturer uses a "mask" to "set" the memory with the correct binary codes. The masked ROM can never be changed. A programmable ROM or (PROM) can be programmed after its manufacture, but generally only once. There is also an erasable programmable ROM, called an EPROM. Its memory can be changed more than once.

Volatile Memory

"Volatile" memory is the section of computer memory that is active as long as the power is on. If the power is turned off, all the binary codes in the RAM (random access memory) are lost. When power to the RAM is reestablished, the registers are all empty but are ready to accept input. The microprocessor writes numbers into the registers of the RAM or reads those data as it processes. Data from sensors may be stored or computed data may be written into these registers. Some of the stored data may be read to make further computations.

Software

A computer program that can be fed into and used by a computer is called *software*. The computer must already have firmware to make use of the software. Any computer program is a sequence of instructions to process incoming data to obtain a specific result. That result may be a letter appearing at a specific position on a CRT screen or the operation of a transister controlling primary current in the ignition circuit of an automobile. The specific computer design determines whether firmware alone or both firmware and software are needed for a given process.

Software programming can be written into the RAM from several different input devices. These could include a keyboard, a disk drive and magnetic disk or a tape drive and magnetic tape. Switches that are part of an automobile's air-conditioning system or other system may also be used to input minimal direction signals to a computer.

Newer automotive computers have or will have keyboards, touch-sensitive CRT screens, or voice-command input systems. Radio signals, microwave signals, and magnetic disk inputs are all part of the future of automotive computers.

AUTOMOTIVE COMPUTERS: A SIMPLE PROGRAM

The object of this discussion is to explain, very briefly, the logic and directions needed within a computer program for it to work. This is not intended to be a lesson in programming. Let's consider the function of the mechanical-advance weights system in the old point-type ignition system. The basic function of that system was to advance ignition timing an appropriate amount as the engine rpm increased. The size of the flyweights and the strength of the flyweight springs determined the ignition advance for a given engine speed.

Now let's look at part of a simulated computer program that attempts the same function but does the job electronically. This attempt is not to write a usable program or even a close representation, but to give you an idea of the direction taken by programmers.

- 02. Start.
- 05. If the rpm is less than 900 and more than 250, read and store the amount of timing advance. (The initial primary firing signal is compared to the TDC input signal.)
- 20. Read the rpm signal. (The buffer has already converted it to a digital code and the firm program has stored it in the proper RAM address.)
- 30. If the RPM is less than 900, add 0° to the existing advance.
- 31. Go to 20. (20 is the program step number. In this case the program is returned to that step again.) This is a test/branch function. The program will continue around this short loop until the rpm exceeds 900. If the rpm exceeds 900, the program will continue to the next step.
- 40. If the rpm is greater than 899 but less than 1150, go to 271. (Here the program is trying to determine what the engine rpm is so that it can assign the proper advance.)
- 50. If the rpm is greater than 1149 but less than 1400, go to 281.
- 60. If the rpm is greater than 1399 but less than 1600, go to 291.
- 70. If the rpm is greater than 1599 but less than 1800, go to 301.

The program steps would continue until the complete rpm range of the engine was programmed. The rpm increments for each test/branch program step can be any values the engineer chooses. Let's pick up the program at the first test/branch step number.

- 271. Determine the initial ignition advance setting. (The computer compares the actual timing of the ignition to TDC by reading the timing of the last ignition firing below 900 rpm. Those data will be stored in a specific register by step 5.)
- 272. Add to the initial ignition timing 1° of advance.
- 273. Go to 20.
- 281. Determine the initial advance setting.
- 282. Add to the initial ignition timing 2° of advance.
- 283. Go to 20.
- 291. Determine the initial advance setting.

You can see that the test/branch function of the program allows the programmer to add a specific amount of advance for each specific range of engine rpm. The program is a series of loops that bring the process back to step 20 after each operation is performed.

The actual program could be more or less complicated depending on the firmware designed into the specific automotive computer. The point of the demonstration above was to see how a computer deals with one variable. Consider the scope of current automotive programs that deal with multiple variables, such as engine temperature, atmospheric pressure, oxygen sensor output, charge pressure, and on and on.

In addition to determining the specific ignition timing needed to gain optimum engine efficiency, the program must deliver optimum emission control, driveability, mileage, and any other demand by the consumer or the government. At the same time the computer is controlling many other functions, such as fuel scheduling, charging system output, air injection to the exhaust system, and idle speed.

Speed is the key to all these functions occurring at what seems to be the same time. Actually, only one step is processed by the computer at any given instant. But the computer is capable of doing many thousands of steps, in correct order, each second. The onboard automotive computer allows designers to control ignition, fuel, emission, and other systems with such accuracy that overall efficiency gains are far beyond our hopes of just a decade ago. The automotive consumer is also enjoying greatly reduced maintenance schedules as part of the computer-controlled automobile.

Automotive service technicians must understand basic computer concepts to provide quality service to customers. As designers work more with this new technology, service will become easier. The technician does, however, have the responsibility to learn how these systems operate and how to service them quickly and safely.

SEVEN

Microprocessor-Controlled Systems

AUTOMOTIVE MICROPROCESSOR USES

The use of microprocessor control in the automobile is increasing at an astounding rate. Virtually every electrical and most mechanical functions of the automobile will be microprocessor controlled in the near future. The major areas now include engine systems management, chassis management, body systems management, and information and entertainment management (Fig. 7-1).

The engine control systems may include emission control, fuel control, ignition control, electrical system control, and communication with other systems. The body systems can include the heating, air conditioning, ventilation, and cruise controls; lighting; seats; windows; entry; and locks. The information system will often include the dash display system, the radio and outside communication, possible navigation aids, and trip-related data such as miles to destination, fuel use, and a host of other quantities. Most of the quantitative information is also available in metric figures.

ENGINE MANAGEMENT

Each of the following engine-related systems is integrated into the general engine computer control system. These systems are given dif-

Chap. 7   Microprocessor — Controlled Systems   61

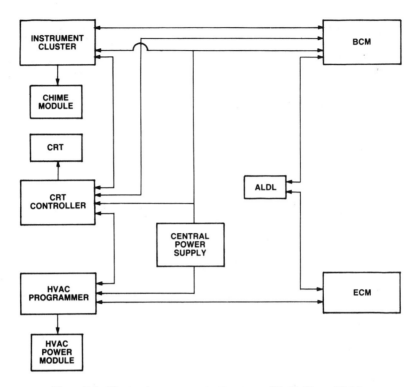

Figure 7-1 Electronic components. Courtesy of Buick Motor Division.

ferent names by various manufacturers but perform very similar jobs (Fig. 7-2). The purpose of this section is to discuss the components in each subsystem and their effect as controlled by the central computer. There are typically six or more engine management sensors that gain and transmit information to the computer. The relationship of this information to each engine control system or related system will be discussed. How most of the currently used sensors actually function is discussed in Chapter 8.

Ignition Spark Timing

Combustion is initiated by the spark plugs to gain maximum fuel economy, lowest emissions, smooth and quiet engine operation, and power required by the driver for the existing conditions. Current to the primary winding of the coil is shut off by computer control of the power transistor in that circuit. On newer engines with multiple coils the control is to the specific coil responsible for firing the next spark plug in the firing order.

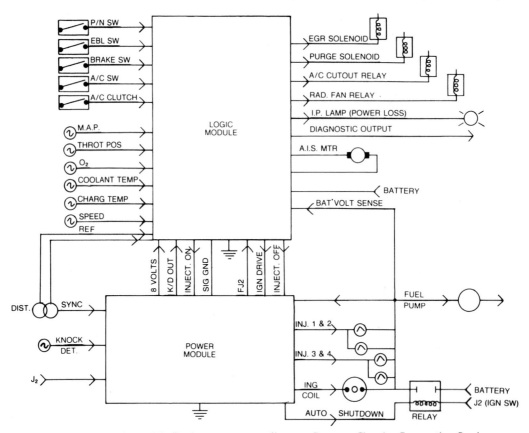

Figure 7-2 Engine management diagram. Courtesy Chrysler Corporation Service Training.

This relationship has become multidimensional with the advent of more complex computer programs. Consider the effects of the varied input of the following sensors. Each one may affect the amount of timing advance or retard of the spark, and in some cases whether the spark plugs fire at all. In sophisticated systems the timing for each cylinder may be determined separately each time it fires.

The engine *crankshaft position* is sensed by a magnetic sensor in the crankcase, at the front pulley or damper, in the distributor, or in some other system. From its electrical output the computer program can determine engine speed and the rate of change of that speed. Higher engine rpm usually requires more advance (Fig. 7-3).

Throttle position can be sensed by a manifold switch, a manifold pressure sensor, a throttle position switch, or a throttle position sensor. Here the specific driver demand and the rate of change of that demand are the quantities that may affect ignition timing. Higher manifold

Chap. 7   Microprocessor — Controlled Systems   63

Figure 7-3 Distributor-type crankshaft-position sensor.

Figure 7-4 Throttle-position sensor.

pressure or a more open throttle usually requires less ignition advance (Fig. 7-4).

Engine temperature may be sensed by a coolant temperature sensor, a cylinder head temperature sensor, an engine oil temperature sensor, or other sensors. Cold engine operation may allow more timing advance than warmer operation. In some cases slightly increased advance is used to help cool excessively hot engines (Fig. 7-5).

Barometric pressure is sensed by an absolute pressure sensor that is usually located in the passenger section of the automobile. It measures the pressure difference between absolute zero and the existing air pressure. Changes in the humidity or the elevation where the auto is at a given moment affect combustion. This sensor sends data to the computer to affect both the ignition and fuel programs. Higher altitudes or thinner air, for example, may require more advance or less fuel (Fig. 7-6).

The *detonator sensor* signals the computer when a specific range of frequencies are emitted by combustion. These frequencies accompany harsh combustion known as *detonation*, which exerts extreme

Figure 7-5 Coolant temperature sensor.

Figure 7-6 Barometric pressure sensor.

forces on engine components that will often damage them. The computer program can retard the ignition timing on a cylinder-by-cylinder basis until detonation stops (Fig. 7-7). This sensor is usually mounted in the cylinder head or intake manifold. As more research is accomplished, more sensors may be included in the engine system together with programming to account for additional variables.

Figure 7-7 Detonation sensor.

Fuel Injectors

Two specific decisions are made concerning fuel injectors. First, the exact timing of when a given fuel injector will open must be determined. Second, the length of time that the injector remains open must be determined. These two determinants set the air/fuel ratio and can affect the quality of combustion by the way the mixture enters the combustion chamber.

The *crankshaft-position sensor* is again in use to allow the computer to determine exact fuel injection initiation time. Some current systems fire half of the port injectors at one time. Then the remaining injectors are fired on the next crankshaft revolution. The fuel remains in a mist just outside the intake valve until that valve is opened. The newer programs are starting to time the specific injector for each cylinder to open in relation to the valve timing of that cylinder.

Manifold pressure sensors, hot-wire airflow sensors, vane- and vortex-type airflow sensors, or many other possible new types of sensors either measure *air mass*, estimate air mass, or measure pressure difference to estimate air mass. These data are used to determine base fuel injection open time.

An *oxygen sensor* determines the amount of unused oxygen in the exhaust gas to help the computer program more accurately trim the

Figure 7-8 Oxygen sensor in the exhaust pipe.

injector time open to maintain the exact 14.7:1 air/fuel ratio. Lack of oxygen in the exhaust gas indicates a rich mixture. The injector time open will be shortened part of a millisecond to lean out the mixture. This sensor is usually located in the exhaust pipe near the engine (Fig. 7-8).

A *temperature sensor* sends data that may require the program to modify the fuel mixture for cold starting or for excessively hot running. Other sensors that might affect the fuel program include a throttle-position sensor or switch, an exhaust temperature sensor at the catalyst, a gear-position switch, and more to come.

Fuel Pump

The fuel pump for most new electronic fuel injection systems is of the electric type and is placed inside the fuel tank. Energy to the fuel pump is controlled by the computer and is affected by several inputs to the program. Power to the computer usually first starts the pump operating. After the first second or two, most programs require a signal from the crankshaft sensor indicating more than 100 rpm or so to maintain power to the pump. This is to keep fuel from being pumped if the engine is not running or being cranked.

Any time the engine stops running, the fuel pump will be turned off by the computer. In the event of an accident, this part of the program is a safety feature. There is also a rollover switch that would signal the computer to shut down the pump if the vehicle were to turn upside down.

Engine Cooling Fan(s)

Most newer vehicles use an electrically driven fan to draw or push air through the radiator. The fan (or fans) is in use only when the coolant temperature exceeds a specified point. This saves engine power wasted turning a fan needlessly and makes control of engine temperature more accurate. Also, with front-wheel drive, a mechanically driven fan is difficult to set up. The coolant temperature sensor is the most important input to the program for this system. Air-conditioning switch data and vehicle speed may also be factors in controlling the fan drive (Fig. 7-9).

Exhaust Gas Recirculation

Exhaust gas recirculation (EGR) is used to curb peak combustion temperatures and reduce the formation of nitrous oxides. Exhaust gas is

Figure 7-9 Electric fan on radiator.

Figure 7-10 EGR valve on Buick. Courtesy of B-O-C General Motors.

allowed to transfer into the intake air supply through a valve interconnecting the two manifolds (Fig. 7-10). The EGR valve is usually operated by diaphragm movement to which modulated vacuum is applied. Power control of the solenoid-operated vacuum modulator is part of the computer program. The result is exhaust gas recirculation control by the computer. Inputs that may affect EGR include engine temperature and the throttle position, with other variables possible.

Canister Purge

The fuel vapor storage canister contains several ounces of charcoal and an air filter. Fuel vapors from the gasoline tank are routed to the storage canister whenever pressure in the tank forces the vapors out. The charcoal holds the vapors until air is drawn through the canister and into the engine to react in the combustion process. A computer-controlled solenoid valve allows manifold vacuum to be applied to the canister under certain conditions (Fig. 7-11).

Inputs that may affect canister purge include engine speed, length of time the engine has been running, whether the engine is operating in closed-loop mode, and engine temperature. Again it is the program with the inputs discussed above that can control the operation of the solenoid valve (Fig. 7-12).

Figure 7-11 Charcoal canister.

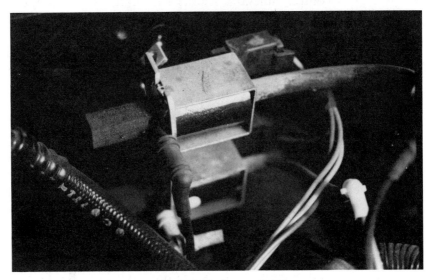

Figure 7-12 Purge solenoid.

Idle Speed Control

Control of the engine idle speed can now be delegated to an idle speed motor. This device changes the idle airflow to gain an exact and constant idle speed. Some engine idle speed control systems may also use fuel mixture to change idle speed over a small rpm range (Fig. 7-13).

Figure 7-13 Idle speed motor.

The crankshaft speed sensor is the main input to the computer for this control function. Additional inputs to the computer may include signals from such systems as air conditioning, electric backlight heater, gearshift position, and various engine controls. In effect, the computer program can anticipate oncoming systems that will affect engine speed and change the throttle position or other controls to keep the engine speed constant.

Diagnostics and Other Functions

The computer program is designed to expect inputs from all the various sensors, switches, and internal units, such as the timer. The input of each of these is checked continuously in a specific order and compared to a range, stored in the computer, to see if the signal is within parameters. If no signal is received or the signal is consistently out of the set parameters, a special code is activated in the diagnostic readout memory section. This code may remain in memory in some systems even if the fault corrects itself. Some programs keep the fault code active for, say, 30 engine starts. Other programs show the fault code only if the malfunction continues to exist.

Data from the vehicle speed sensor, the throttle-position sensor, or the manifold pressure can be used to control the transmission shift points and the torque converter clutch. The data collected from various sensors can also be sent on through the communication network of the car to provide information to the driver or the mechanic for diagnosis.

The charging system can also be controlled by the engine management program of the computer. The voltage level of the generating system is sensed as it is input to the computer. Energy to the rotor of the alternator is continuously controlled within the computer. Very accurate voltage levels can be maintained, yet eliminate the separate voltage regulator now used by many manufacturers.

BODY MANAGEMENT

The functions being added to the body management computing system are a very fast rising collection. The inputs to these systems include switches, sensors, and input from touch-activated CRT circuitry. Voice inputs are a possibility for the future (Fig. 7-14).

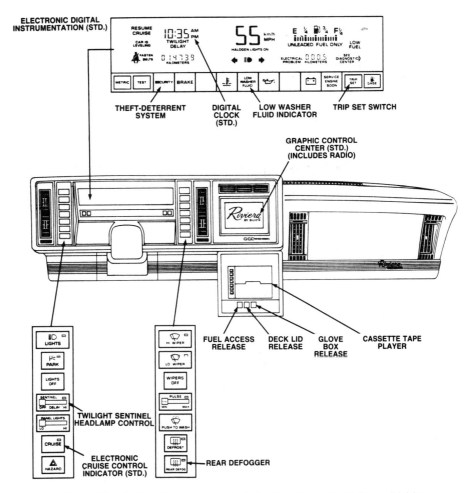

Figure 7-14 Buick Rivera instrument cluster. Courtesy of Buick Motor Division.

Heating, ventilation, and air conditioning was the first major body system to receive computer control. A typical system now receives a specific temperature input from the driver. The computer then maintains that temperature within the passenger area using either heating or air conditioning, as needed. The level of air change, the fan speeds, or the humidity level are programmable in some systems. Sensors might include multilevel temperature sensors, humidity sensors, and possibly photosensors to determine visual conditions of the windshield and backlight. Switch inputs for doors, seats, safety belts, and other systems are also fed to the computer.

Cruise control is another microprocessor-controlled function. The location of this function may be in the body management system computer or a separate processor receiving inputs from the proper sensors. The vehicle speed sensor, brake pedal switch, clutch pedal switch, transmission gear selector switch, or engine speed sensor may be used to provide inputs to this system.

Vehicle handling control systems include suspension stiffness, automatic brake control, and self-leveling. Sensors used in these systems can measure wheel speed, suspension height at each suspension member or lateral or vertical acceleration, and body attitude. Inputs from these sensors are received by the computer control system and activate the proper valves to pulse brake pressure, increase air pressure to shock absorbers, or change the fluid flow rate through shock absorber control orifices. A hydraulic or pneumatic pump, reservoir, valves, and linear actuators may also be included in these systems. Driver input as to the type of handling desired or other factors is also received. Switches, a keyboard, or touch-sensitive CRT can all be used as input devices.

The body control system also gathers information and sends data on to the instrumentation control system as well as the engine control system. The voltage to the lighting systems can be monitored so that burned out lights may be noted and displayed on the driver information system.

All of these inputs can be measured against programmed standards or ranges for continuous monitoring. Improper data are then used for diagnostics and the program then can set a fault code in the body diagnostics memory section for driver or mechanic repair information.

INSTRUMENTATION DISPLAYS

The continual changes in the instrumentation display systems are the most noticeable part of the influx of computer-controlled automotive systems. These instrumentation displays are also one of the best salesmen of new automobiles and carmakers are taking full advantage of that fact. On several new autos, virtually the entire inside of the car lights up as you sit down and close the door.

Chapters Four and Five dealt with the operational theory of the specific devices used to display the information. There are four general types of display systems that use the different display devices, including segment and dot matrix, VF telltale panels, CRT, and incandescent telltale panels.

The multipoint displays could be in the seven- or nine-segment pattern for numeric information only, or could be designed with 20 or more dots in the matrix pattern for numbers, letters, or even limited graphics, The three major types of devices used at this time are vacuum fluorescent in several different colors, light-emitting diodes which are also becoming available in several different colors, and liquid-crystal display using colored backlighting.

The telltale panel system is the old standby using a printed message on a colored surface. When the panel is backlighted, the message stands out. This system has been used for the last 30 years in the form that used to be called "idiot lights." That started when the warnings were first substituted for mechanical gauges, such as oil pressure, charging system, and coolant temperature. Incandescent bulbs were used for most of that time. Now the trend is to use a vacuum fluorescent lighting system with the information panel.

The newest display system for automobiles and most likely to become prominent is the cathode ray tube (CRT). The information we now receive on our home television set is an everyday part of our lives. It will be a natural step to accept information from a dashboard display CRT in an automobile. One of the major advantages of this system is that the type of information that can be displayed is almost limitless. The second major advantage is the use of the touch-screen concept to input choices or data into the system. Vehicles now featuring the CRT use it for other than the primary functions of speed and engine conditions (Fig. 7-15).

The CRT is now also being tested as part of an interactive navigation system. Information could be input to the navigation display by disk, tape, radio-wave, or microwave transmission. The automobile's immediate position could be shown on the most current maps of the area. Possible route information could be scanned. The nearest commercial, historical, or recreational sites could be listed as you travel through different areas. This is just the tip of the iceberg in providing the traveler with useful information.

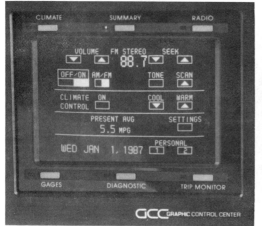 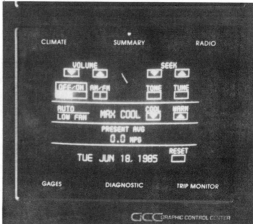

Figure 7-15 CRT summary. Courtesy of Buick Motor Division.

Voice and Sound Alert Systems

We have become used to listening to the seat-belt reminder buzzer or chime. Sound alert systems are a part of automobiles produced since the 1970s. What is new is the number of system sound alerts that are now computer controlled.

The chime alert (or buzzer on less sophisticated vehicles) is programmed to signal that lights are left on, doors are not shut, fuel is low, or any of several dozen other possible conditions that the engineers feel need to be signaled. The single tone may be used at a constant level or pulsed at either fast or slow rates. Different tones can be used for different warning functions also. The warning tones are also used in conjunction with telltale panel indicators. In the case of the seat-belt warning, there is usually a timer that keeps the circuit active for only 8 to 10 seconds.

Voice-alert modules are in use for some automotive warning systems. The sound is generated by an IC chip and emitted by the speaker. Input from various switches and sensors can be used to trigger this particular computer program response. At this point the use of voice alert is more of a sales tool than a serious interaction with the driver and the auto.

More sophisticated programs could provide a higher level of usefulness for the voice-alert system. It would be particularly useful where the driver should not take his or her eyes from the road. The voice-alert system also provides a second communication system so that the visual mode is not overwhelming.

Theft Deterrent and Driver Entry

The automotive accessory aftermarket area has had computer-controlled theft-deterrent systems available for years. Motion sensors, hood open switches, and other sensing devices are used to sense when people other than the owner are tampering with the car. Buzzers, horns, and the use of radio transmissions to a remote receiver can all be part of the system. In the near future more sophisticated systems will be manufacturer options. Currently, ignition keys with different electrical resistances are used in the Corvette to deter theft. If a wrong resistance is sensed, the auto will not start.

Data input by pushbutton for keyless entry has also been used. Other computer-sensed systems can be used to turn off alarms or open doors. Driver-response measurement systems have been suggested to keep inebriated or otherwise impaired drivers from starting the vehicle. This is an area for anticipated expansion.

The systems discussed are only some of the electronic systems that we will become used to in the next decade. New sensors and other input data systems will be devised. More output modes will also be created. We are just at the beginning of microcomputer-controlled automotive systems.

EIGHT

Automotive Sensor Technology

ENGINE SYSTEM SENSORS

In this first section we describe and illustrate many of the common sensors used on automotive engines or relating to their operation. Some sensor technology is rather old and commonplace, while many new sensor designs are just appearing. We will certainly see more new sensor designs in the near future.

An interesting aspect of sensor design is the environment within which the sensor must exist. Consider the effect on simple sensor design of engine vibration, outright physical shock, temperatures ranging from -40° to 275°F, thermal shock, various mechanics, dirt, salt, snow, and a full range of humidity. Each sensor concept must incorporate the solution to these potential problems as well as the design of the actual sensor.

Another problem encountered is the designed range over which the sensor must sense or signal. Engine displacement, fuel system differences, or many other considerations must be accounted for in the initial design stages. A large manufacturing company wants to avoid the use of several sensor sizes to cover a range of engine sizes. The sensor design must be capable of that flexibility.

Throttle Position

A design used for this function uses the change in resistance of an electric coil as the variable (Fig. 8-1). A simplified concept is to think of

Chap. 8　Automotive Sensor Technology

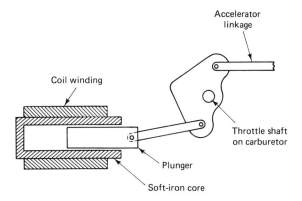

Figure 8-1　Reluctance change sensor: plunger and coil for throttle position.

the resistance of the coil changing as the plunger part of the sensor is moved into the hole in the center of the sensor coil. As the resistance increases, the voltage decreases. The signal sent to the computer is an increasing or decreasing voltage. The computer program sees each specific voltage as a given throttle position. Both the throttle position and the rate of change of throttle position are measured from the output of this sensor.

The sensor may be checked for continuity and correct resistance with a sensitive ohmmeter. Using a digital voltmeter with a minimum internal resistance of 10 megohms, you can check the sensor for voltage changes while it is connected and operating.

Oxygen Sensor

The zirconia oxygen sensor is in effect an electrical generator that produces a small voltage under the right conditions (Fig 8-2). The direction of the potential of the electrical output changes as the exhaust gas changes from excessive oxygen to excessive fuel. In effect, the sensor acts like a switch going off and on as the voltage changes from positive to negative.

The core of the sensor consists of a hollow ceramic bulb or tube-like structure coated with a platinum film and a protective coating. Surrounding that is a metal shield with perforations to allow exhaust gases to come in contact with the bulb. The inside of the bulb is vented to the atmosphere. The difference in oxygen in the gases on the two sides of this particular molecular structure, within a given relatively high temperature range, will create the electrical potential that becomes the signal to the computer.

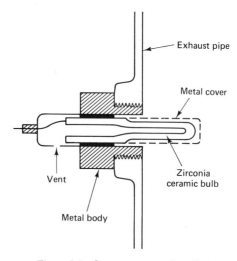

Figure 8-2 Oxygen sensor: zirconia.

This signal will switch from plus to minus or back as the exhaust gas flow crosses the stoichometric air/fuel ratio. Testing the sensor with a digital voltmeter under dynamic conditions shows a voltage with changing polarity or a voltage ranging from 0 to above 500 millivolts. Higher voltage indicates a rich (lack of oxygen) mixture.

Crankshaft Position

This most important sensor is found in many different configurations. The use of some form of magnetic field is common to many of the sensors. Of the more common types, the reluctance sensor is in wide use (Fig. 8.3). It consists of a permanent magnet with a coil surrounding it.

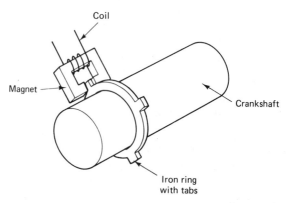

Figure 8-3 Crankshaft position sensor: reluctance type.

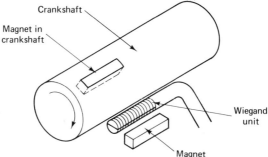

Figure 8-4 Wiegand effect crankshaft-position sensor with crankshaft coil and magnets.

A metal tab passing close to the magnet fluxes the magnetic field across the coil, which in turn causes a change in the reluctance of the coil. A current being sent through the coil would change. The momentary change in the current is the output signal of the sensor.

The *Wiegand effect sensor* consists of a special type of wire in a coil with a magnet behind it (Fig. 8-4). A second magnet is placed in the crankshaft in position to rotate past the coil. A voltage is generated in the coil in a specifically crisp format each time the crankshaft magnet passes. The special magnetic wire of the coil causes a narrow burst of voltage rather than a slow buildup and decline.

The *Hall effect sensor* uses a semiconductor device installed in a magnetic field (Fig. 8-5). There is a metal tab on the rotating shaft that when passing the magnet, causes the magnetic field to flux or move about somewhat. The normal current through the semiconductor is disturbed by the magnetic fluxing and momentarily generates a reverse current, which is the signal set out by this sensor. Drawbacks of this

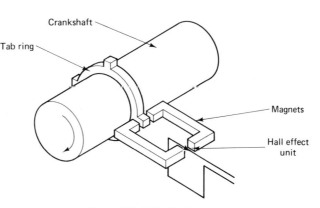

Figure 8-5 Hall effect sensor.

sensor are the semiconductor's sensitivity to heat and its low current output.

Less commonly used crankshaft-position sensors include magnetic reed switch devices, photo-optical sensors, and variable-inductance sensors. These have less desirable traits at this stage of development and will not be explained.

Manifold Absolute Pressure Sensor

The measurement of the pressure difference between atmospheric and intake manifold pressure is one of the basic ways that mass airflow can be approximated. Electronic fuel injection needs some fast-responding airflow measurement to determine the correct portion of fuel to inject to obtain the proper air/fuel ratio. Several designs have been developed and put into production.

The aneroid linear variable differential transformer (aneroid/LVDT) is one of the earlier designs (Fig. 8-6). Two or more metal diaphragms are connected together with an evacuated internal area. A rod attached to the transformer core is fastened to one side of one dia-

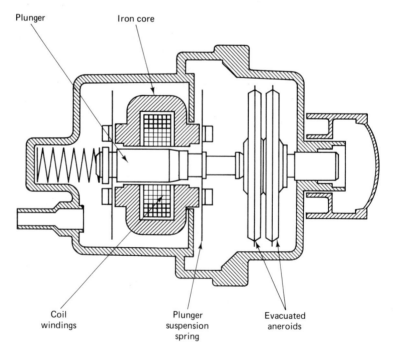

Figure 8-6 Aneroid linear variable differential transformer. Courtesy of Bosch.

phragm, the other side being attached to the device's case. Manifold pressure/vacuum is applied to the outer area of the aneroid bellows, causing it to expand or contract, depending on that variable. The bellows movement is transferred to the transformer core.

The transformer part of the sensor consists of two coils, an input and an output. A constant-voltage ac is sent to the input coil, but the output varies as the core is moved into and out of the coils. This variable output is the sensor signal. This sensor was common to the earlier Bosch EFI-D fuel injection systems. Pressure sensors may be absolute with a fixed pressure reference, or differential, measuring between two pressures.

The capacitor-capsule manifold absolute pressure (MAP) sensor is now one of the main formats used (Fig. 8-7). The structure consists of two very thin ceramic-based disks with a spacer, the whole assembly being fused together. The internal area is evacuated. The inside surface of each disk is coated with a conducting film that is electrically connected to outside leads. In effect, an aneroid exists, with the inside surfaces that face each other comprising a capacitor. The capacitance is inversely proportional to the distance separating the two inner surfaces. An electronic support circuit is necessary to make use of this type of signal.

A strain-gauge pressure sensor is also manufactured to measure MAP (Fig. 8-8). The basic configuration of this sensor is a tiny thin silicone disk with several resistors deposited in series just in from the edge by microelectronic production techniques. The disk is supported at the edges and deflects from exposure to pressure differences. The deflection changes the resistance of the circuit, which is the sensor signal. This design is adaptable to oil pressure and other pressure-sensing data collection.

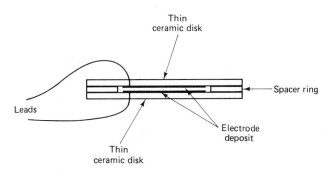

Figure 8-7 Capacitor-capsule MAP sensor.

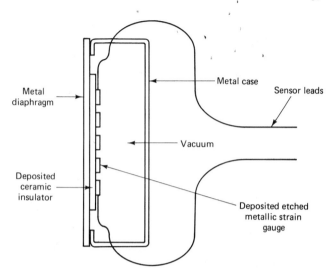

Figure 8-8 Strain gauge pressure sensor.

Another diaphragm, much larger than the capacitor capsule type, is the variable inductor diaphragm (Fig. 8-9). This design is particularly good for manifold vacuum readings. The parts consist of a coil wound over a hollow core, a core plunger with an attaching stem, and a diaphragm that is spring loaded. The diaphragm with its spring loading is set up with atmospheric pressure on one side and manifold vacuum on the other. As the vacuum changes, the plunger is moved in and out of the coil, changing its inductance. Other designs are beginning to be used and will be included as they warrant recognition.

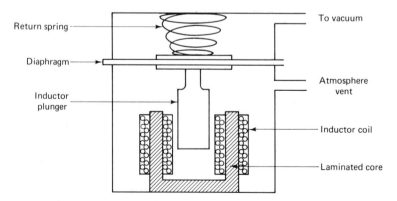

Figure 8-9 Variable inductor diaphragm sensor.

Knock Sensor

Knock, as a deviant form of combustion in an automotive engine, occurs within a specific frequency range. The design of the knock sensor must be sensitive to these frequencies and exclude all others. The magnetostrictive knock sensor consists of two rods of a specific length centered in an electric coil. The assembly also contains a magnet and spring (Fig. 8-10). Engine vibration of the specific frequencies causes the rods to vibrate, which in turn changes the magnetic field around the coil. This change, as it affects the coil current, produces the sensor signal.

Temperature Sensor

A number of designs are being used to sense both coolant and air temperature. The coolant temperature sensor must survive in a somewhat harsher environment. The air-temperature sensor must have a faster response time. In both cases the response should be planar or parallel in output to the change in temperature. Temperature sensors are also finding use sensing engine oil temperature and gearbox oil temperature.

The semiconductor resistor temperature sensor is a semiconductor material that is sensitive to temperature changes. DC current through the material lessens as the temperature increases. The change in current is the sensor signal (Fig 8-11).

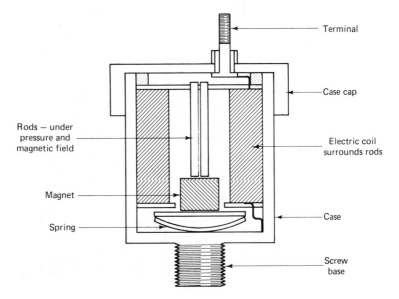

Figure 8-10 Magnetostrictive knock sensor.

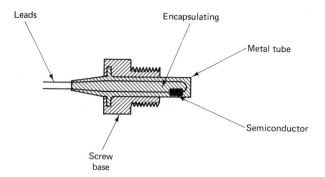

Figure 8-11 Semiconductor resistor temperature sensor.

The thermistor is also a solid-state device used to sense variable temperature changes. A change in resistance occurs that is planar for the temperature change. Both fuel and coolant levels are also now monitored for transgression below a safe level using thermistor circuity. In the case of the coolant level the thermistor is inserted partway down the radiator (Fig. 8-12). When the coolant volume gets lower than the level of the thermistor, a change in the resistance triggers a warning light on the instrument panel.

The wire-wound resistor temperature sensor consists of a length of moderate-resistance wire wound around a stable core (Fig. 8-13). The resistor is placed in material suitable for insertion into the coolant system. The resistor is sensitive to the changes in temperature of the coolant because its electrical resistance varies fairly linearly to that temperature change.

Figure 8-12 Thermister temperature sensor in the radiator.

Chap. 8   Automotive Sensor Technology

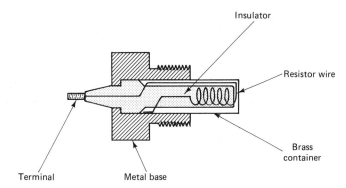

Figure 8-13   Wire-wound resistor-type temperature sensor.

Airflow

Two major approaches are used to determine the appropriate base fuel metering with respect to the air taken in by a fuel injection engine: air pressure difference, and airflow measurement. The speed/density system (pressure measurement) lags changes in the airflow rate compared to the mass-airflow system (vane or hot-wire). The early Corvettes used a special venturi to produce a strong vacuum signal. Later, Bosch electronic injection first used manifold vacuum and then an airflow measurement system.

The vane or flapper door airflow measurement consists of a lightly spring loaded valve that moves aside as airflow increases (Fig. 8-14).

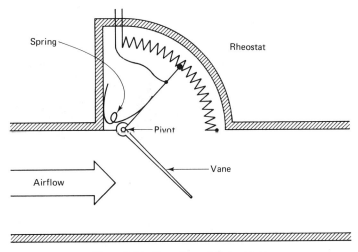

Figure 8-14   Vane-type airflow sensor.

The valve is tied to a rheostat, a type of variable resistor. The change in current in the resistor circuit is the sensor signal. Also used is a carbon-film resistor with variable area connected to the airflow meter plate. It gives a signal that varies air/fuel ratio with demand.

A newer type of airflow meter uses a hot wire to sense airflow. The wire is stretched across a specially designed intake section of the engine and heated by sending an appropriate current through it (Fig. 8-15). As airflow increases, cooling the wire, resistance decreases. A bridge circuit senses the decrease and increases the current, which brings the wire back up to its original resistance and temperature. The increased current is the sensor's signal. There is also an air-temperature compensation element. Newer designs could incorporate the wire being wound in a coil and placed in a bypass channel to protect the system better.

Another airflow sensor uses the effect of air turbulence on ultrasonic waves to measure airflow (Fig. 8-16). A specially shaped cone put in the air intake path produces vortices in proportion to the air velocity through the passage. The ultrasonic sender and receiver are on opposite sides of the passage and the vortices lower the air density in that area. This change in the receiving signal, after processing and amplification, becomes the sensor output signal.

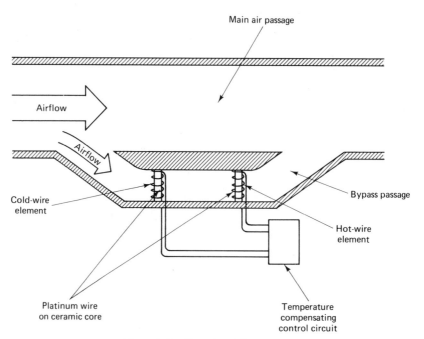

Figure 8-15 Hot-wire airflow sensor.

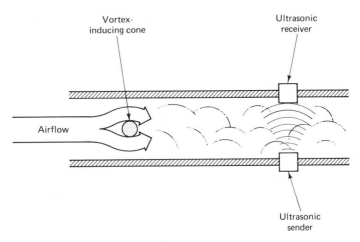

Figure 8-16 Ultrasonic airflow sensor.

Fuel Flow

One simple fuel-flow measurement system in electronic fuel injection applications is to use injector-open-time values. If the injector orifices are accurately controlled and the fuel pressure is constant, the injector open time is closely proportional to fuel flow. Adding up the open time of all the injectors involved becomes the sensor's signal.

The ball-in-race fuel-flow sensor can use an optical system to pick off the fuel-flow rate. Fuel is directed through the sensor and forces a ball to travel around a circular race or path (Fig. 8-17). Each time the ball passes a light source, usually a LED, that light source is interrupted.

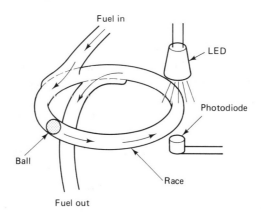

Figure 8-17 Ball-in-race fuel-flow sensor.

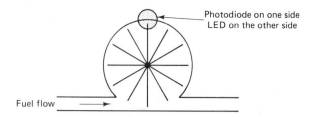

Figure 8-18 Paddlewheel fuel-flow sensor.

The light-receiving photodiode then sends a pulse to the computer as the flow rate signal.

Another fuel-flow sensor design incorporates a paddlewheel. Again an LED and a photodiode are used to react to the passing of each paddlewheel spoke (Fig. 8-18). Increased fuel flow turns the paddlewheel more rapidly and increases the output signal of the sensor.

AUTOMOTIVE SENSORS

Wheel Speed

Metal protrusions on the outer circumference of the brake disk passing close to a magnetically fluxed coil will cause a change in current through the coil which is caused by the change in reluctance of the coil. This current change is the sensor's signal.

Humidity

Humidity within the passenger compartment can be sensed and when necessary, can signal the climate control system to operate the ac system to bring the humidity level within the passenger comfort range.

The salt-doped semiconductor humidity sensor is of quite simple design. The sensor surface is a semiconductor connected into the circuit from front to back. The surface is doped with a special salt that has the property of increasing the conductivity of the semiconductor with the higher presence of water in the air (Fig. 8-19). The H_2O is then changed to H_3O and that ion carries the current into the semiconductor material. The increased current is the sensor's signal.

The capacitor-type humidity sensor has a semiconductor surface that has a vertical crystalline structure (Fig. 8-20). The presence of water attracted to the porous surface changes the capacitance of the semiconductor. That change is the sensor's signal.

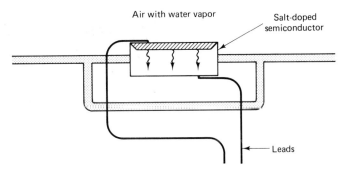

Figure 8-19 Salt-doped humidity sensor.

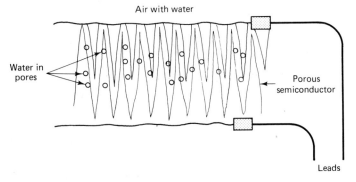

Figure 8-20 Capacity-type humidity sensor.

Vehicle Speed

The vehicle speed sensor found in many domestic automobiles is a permanent-magnet generator (Fig. 8-21). The sensor is often gear driven, similar to the way that speedometer cables used to be. The gear turns a small shaft-mounted permanent magnet. The magnet rotates inside a wire-wound coil.

That rotation produces a sine-wave current proportional in voltage to the magnet's rotation speed. An electronic buffer converts the signal to a square-wave form that can be interpreted by the microprocessor. Some earlier speed sensors were also mounted in line in the speedometer cable.

Suspension Height

One way that suspension height can be sensed is by using a linear transducer that converts lineal suspension travel into a digital electronic sig-

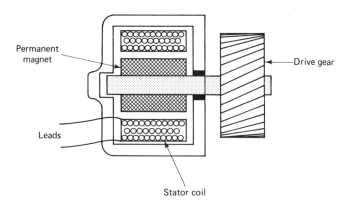

Figure 8-21 Permanent-magnet vehicle speed sensor.

nal. The use of Hall effect switches can produce that signal. To understand the Hall effect, consider a conductor that is carrying current from point A to point B (Fig. 8-22). When a magnetic field is applied to the conductor, voltage is produced across the conductor from point X to point Y. That voltage is the Hall effect.

In the suspension height sensor two Hall effect switches are positioned so that a shaft with two permanent magnets attached, of opposite polarity, will pass next to them (Fig. 8-23). The shaft is attached to the suspension so that the magnets are slid up and down past the switches as the relationship between the car body and the suspension height changes. The output Hall voltage of the switches, as affected by magnets of opposite polarity, is the signal to be interpreted by the logic of the microcircuitry.

Suspension height may also be signaled by variable resistance. A lever arm attached to a movable suspension member can drive a contact across a wire-wound or thick-film variable resistor (Fig. 8-24). The change in resistance is the sensor's output.

Steering Position

An optical sensing system handles steering-position regulation well. Two LEDs and two phototransistors are positioned opposite each other.

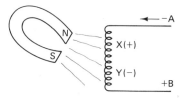

Figure 8-22 Hall effect switches in suspension point A to B–X to Y.

Chap. 8   Automotive Sensor Technology

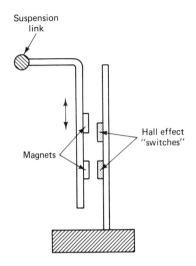

Figure 8-23   Actual hall effect suspension setup.

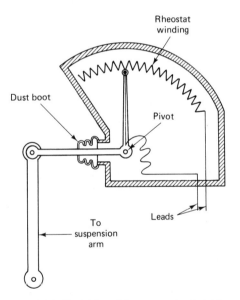

Figure 8-24   Variable resistor suspension height sensor.

A slotted disk attached to the steering shaft rotates between the semiconductors and interrupts the signal being sent and received (Fig. 8-25).

The location of the two LED/phototransistor sets is spaced so that a phase difference in their output signal is present when the steering shaft rotates. Essentially, the distance between them is different from the spacing of the slots in the disk.

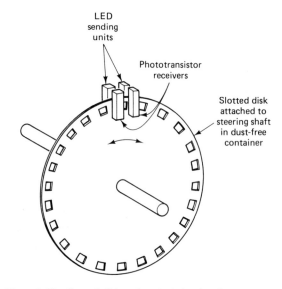

Figure 8-25 Slotted disk and optical signal and sensor system.

Fuel Level

The old standby fuel-level sending unit consists of a wire-wound float-lever-activated rheostat (Fig. 8-26). As the fuel level descends, the float and its sliding contact move down. This causes a change in the resistance of the sending unit. That change in resistance is the sending unit's signal.

A newer technology uses a thick-film deposit on a plastic strip to produce the change in resistance for the sensor signal. The film is placed directly in the fuel. The level of the fuel on the film changes the film's deposit resistance. The shape of the deposit determines if the output will be used in a digital or an analog display system (Figs. 8-27 and 8-28).

Switch Signals

The signal required for many microcomputer decisions or outputs is a simple yes or no, voltage or no voltage, on or off. A simple switch is the sensor. The major purpose of the switch might be to turn on the interior light when the car door is opened. That same signal can at the same time be used to inform the driver that a door is still open.

Switches are used to signal seat-belt connection, seat occupied, ac on to the idle control system, rear backlight heater on, and so forth.

Chap. 8 Automotive Sensor Technology

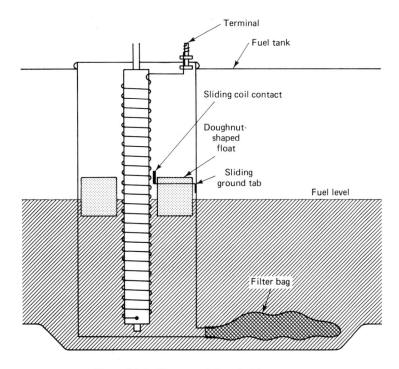

Figure 8-26 Wire-wound-float fuel-level sensor.

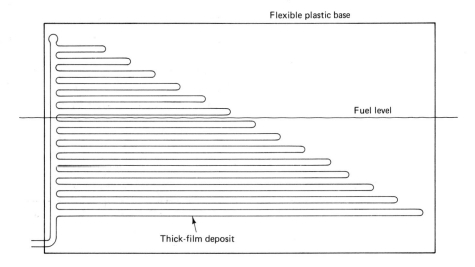

Figure 8-27 Digital thick-film fuel-level sensor.

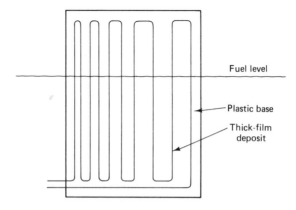

Figure 8-28 Analog thick-film fuel-level sensor.

Many switches now report to the computer control systems as well as performing the current control function for which they were originally designed.

SENSOR BUFFERS

The signal that many sensors produce is a varying voltage or an analog signal. The computer does not know what to do with that type of signal. The signal must be converted to a digital signal for proper assimilation by the ECU. A *buffer* does that job (Fig. 8-29). The buffer is an electronic device that turns on and off at specific voltage levels as the input voltage varies. The input voltage is analog and the output is digital (Fig. 8-30).

The buffer may also have an amplification circuit to ensure that the output signal is appropriate for the ECU. Buffers must be designed for each specific sensor output circuit for which they are needed.

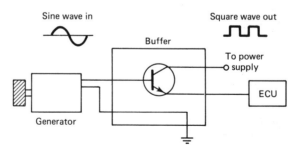

Figure 8-29 Buffer circuit between sensor and ECU.

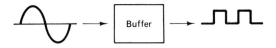

Figure 8-30 Analog input–buffer–digital output.

There is no doubt that sensor technology will become increasingly more important, more complex, and certainly more varied. Automechanics and automotive teachers, as well as the consumer, will need to keep informed in this important area of technology.

NINE

Control Technology

FUNCTION OF ACTUATORS

The microcomputer is capable of receiving and sending signals. Sensors and/or their buffer circuits send the signals that the computer receives. Actuators receive the signals that computers send that need to be converted into action. The action most often takes the form of mechanical force operating through some distance. Other forms of action might include a more powerful electrical response than the computer can directly put out. A chemical response may also be appropriate.

Most automotive actuators are electrical and require power beyond the capability of typical microcomputer circuitry. This leads in two directions. First, the initial output can be used to control some form of electrical power circuit. Second, the output can be used to control a pilot valve for a hydraulic, pneumatic, or vacuum circuit. Within the electronic area, transistorized amplification circuits can be used to amplify the output power of the computer signal.

TYPES OF ACTUATORS

Relays

The relay is a simple mechanical form of electrical power amplification circuit. The basis is an electromagnet attracting a piece of soft iron. The soft-iron armature, the part that moves, is usually spring loaded to keep it back away from the electromagnet when the current is not applied. A spring-loaded armature is, however, positioned very near the electromagnetic coil (Fig. 9-1). Power-circuit contact points are placed both on the armature and in close proximity to make contact when the magnetic coil is energized. The very small current required to power the armature circuit therefore controls the much larger power circuit.

Solenoids

Solenoids are also electromagnetic devices. Their purpose is usually to cause a linear mechanical motion. The construction consists of an insulated wire coil wound over a laminated iron core and an iron armature. The armature is usually spring loaded to hold it a distance away from the coil/core assembly. When the coil is electrically energized, a strong magnetic field is formed, pulling the armature toward it. With the current off and the magnetic field gone, the spring pushes the armature back away from the core/winding assembly.

There are a considerable number of solenoid shapes possible, but the basic intent is to deliver a linear motion with an electrical control. A round hollow-core solenoid draws the round soft-iron armature into it when energized (Fig. 9-2). A C- or E-core-design solenoid will produce a larger force over a shorter distance than will the hollow-core design. (Fig. 9-3).

Torque Motor

A solenoid that is designed to provide a motion in two opposite directions with a "no current" center point can be constructed using two

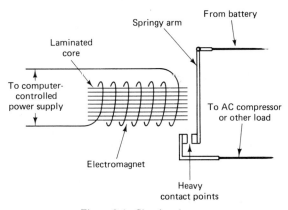

Figure 9-1 Simple relay.

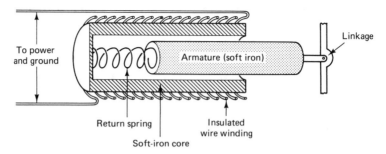

Figure 9-2 Hollow-core solenoid.

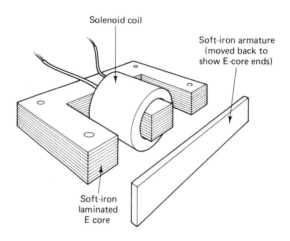

Figure 9-3 E-core solenoid.

coils or with a permanent magnet and current reversal (Fig. 9-4). Energizing either coil pivots the armature on its pinned axis. The opposite end of the armature can be used to control a three-position spool valve or flapper valve.

Solenoid Valves

Solenoid valves for fuel shutoff have been in use in automotive applications for a long time. The introduction of solenoid-controlled fuel injection valves issued in large-quantity production of automotive application solenoid valves. The valve's fast response, relatively low cost, and reliability are indications that the automotive industry will be using them for a long time.

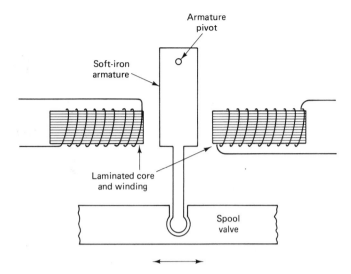

Figure 9-4 Lineal two-direction torquemotor.

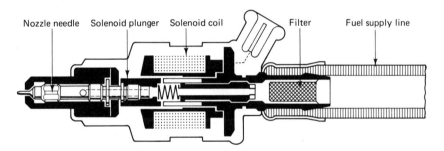

Figure 9-5 Bosch fuel injector: cross section with labels.

The Bosch injector for multipoint injection incorporates a needle valve armature that is spring loaded in the closed position and a copper wire-wound coil (Fig. 9-5).

Stepper Motors

This type of actuator is a current-pulsed rotor and stator that are controlled in the amount of rotation allowed. The rotation control may be a type of ratchet device that allows a set number of degrees of rotation for each torque-producing pulse. The power pulses to the motor are controlled and can be counted by the computer to keep track of the angular movement of the motor (Fig. 9-6).

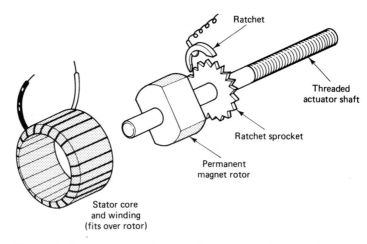

Figure 9-6 Stepper motor with ratchet device to control rotational degrees.

Stepper motors can be used to set idle speed to open and close valves a proportional amount, or to move a lead screw to position various controls. Varying the control rate of shock absorber valves is one application in production.

Thermal Actuators

The bimetallic strip with thermal actuation has been used for many years to operate directional flasher systems (Fig. 9-7). Movement of a

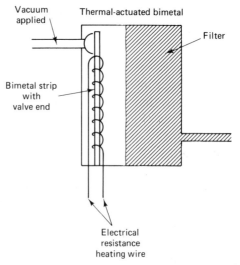

Figure 9-7 Thermal actuator with resistance heater.

heated spiral bimetal strip has also been used to control choke action. Current uses might include control of vapor-pressure valving or movement of heater and ac system components.

The list of actuators at this point is not excessively long, but is certain to increase over the coming years. It is likely that most gains in the development of actuators will be in the electromechanical realm.

TEN

Electronic Service

SERVICING ELECTRONIC COMPONENTS

The question of how to service all the different electronic components has probably crossed your mind more than once while reading the preceding chapters. If you are an automechanic, service is the skill that you get paid for. The type of service that needs to be accomplished in the area of automotive electronics is, however, somewhat different from the conventional mechanical skills most mechanics learned or were originally taught.

Differences exist in the test equipment used, the specific workshop manual sections used in the electronics area, certainly in the knowledge required to be successful, and most important of all, in the attitude and approach the technician must use to be successful in this emerging area. You're going to see *attitude* stressed a lot in this chapter. If the mechanic's attitude isn't right, the costs are going to be high for someone.

EQUIPMENT NEEDED

Let's start with the basic meter you will need to determine voltage, resistance, or current. A standard analog VOM may get you in trouble very fast. It can add or draw excessive current in delicate electronic circuits. What you need is a digital volt/ohm meter with a minimum internal resistance of 10 million ohms. The 10-megohm resistance is needed so that virtually no current will be drawn from or put through any electronic circuit you test.

A meter such as the Fluke model 77 is a good choice. Most measurements are on autoranging scales. The meter decides how big or small the value being measured is. You do not have to worry about the size of the measured value or on some scales, even the polarity, with this type of meter (Fig. 10-1).

The digital readout has two distinct values. First, you will not make many mistakes as to the exact value being measured. The digital scale specifies the number and the value range quite distinctly. Second, the polarity is also very distinctly displayed. On the better meters you can "freeze" the reading so that an instantaneous decision on the value also does not have to be made (Fig. 10-2).

The scales you will use the most are the dc voltage and the resistance measurement or the ohms value. On the meter scale selector

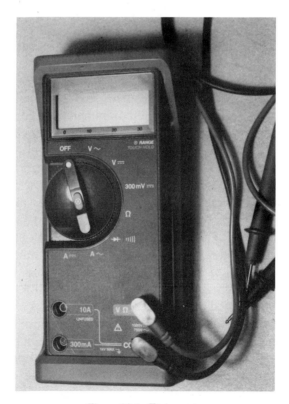

Figure 10-1 Fluke meter.

Figure 10-2 Freeze value on a Fluke meter.

shown, the ____ line with the - - - - line above it represents dc voltage. The Greek letter Ω is used to represent ohms (Fig. 10-3). This meter also has a separate test lead plug-in socket which will measure up to 10 A. Occasionally, this might be useful to an automechanic, but seldom when testing electronic components. A good test meter represents the most important single test equipment item that you will need.

There are dozens and perhaps hundreds of special-purpose electronic test instruments available for use in specific automotive systems. If you are a mechanic who works on just one make of automobile and do a lot of electronic work, you may find the cost of one of these specialized test intruments justified. But there are two problems; these devices are usually limited to a few very specific tests and each year's crop of new cars seems to need a new version of the test device to test all the latest developments.

Another trend in the automobile electronic component area is for the equipment installed in the car to test itself: *self-diagnostics*. The manufacturers are trying to make the need for extra test equipment for their specific product unnecessary. The manual or specific booklet describing the meaning of the diagnostic codes will soon be all that is required beyond the basic VOM. By pushing one or two buttons or touching a space on the CRT you can get a full diagnostic code readout. The future will probably bring a complete verbal description of the fault on the CRT screen and not even require the need for fault codes.

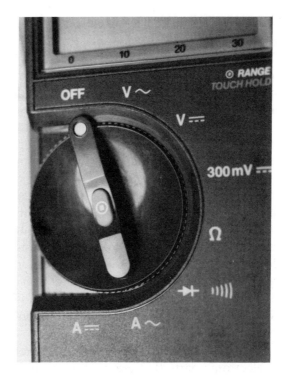

Figure 10-3 Close-up of the selector scale on a Fluke model 77 meter.

SERVICE PHILOSOPHY

Many of us have been able to figure out mechanical systems by taking them apart and understanding the relationship between the various parts. That certainly will not work for electronic circuitry. You have to start with a new and different philosophy. "Philosophy" is used here to mean a way to understand the truth of how something works—a system of studying what you have to learn to be able to diagnose and repair electronic systems.

1. You do have to understand the basic concepts of how an electronic system functions.
2. You have to search out the needed information on the specific system you are going to try to diagnose: know exactly what parts are in that particular system, where they might be located, and what specific value you can test for.
3. You have to think carefully about what the results of your testing means.
4. You have to be committed to doing the very best work possible. Any haste or carelessness may result in a nonfunctioning system.

SPECIFIC SERVICE EXAMPLES

There are some general service points that you should consider before you actually start work on a given system. First, when a system is in operation there may be magnetic fields built up around coil windings or electrical energy stored in capacitors. In some cases if you disconnect a component, with the system operating, to determine the effect on the rest of the system, an electrical energy surge may occur in the remaining operating system that could damage a microcircuit.

Second, electronic circuits that are not self-testing, that is, do not have a built-in diagnostic program, are usually diagnosed as nonfunctioning by eliminating every other possibility first. If the problem still exists, a new electronic circuit is substituted for the original to see if the system will then operate. The antithesis of this is the common practice in the industry that *if you purchase anything electric or electronic, there are no returns.* You cannot take it back and have your money returned. You have to be sure that no other possibility exists before you make the substitution! These components will generally be expensive.

A third situation that is rapidly developing is the large inventory of electronic parts that are used in specific systems. It is of prime importance to get only the part that has the correct part number. Use the utmost care to obtain the exact part number before making any purchase.

Another precaution that needs to be exercised relates to the way in which electronic components are handled. Static electricity can deprogram or more accurately, scatter the effective program in many electronic chips. These chips are shipped in special static-grounding plastic bags. Also, there are often special holders supplied to pick up and insert the chips into their proper connectors. *Do not* touch these chips with your fingers. *Do* use the special holders to insert the components into their mating connectors (Fig. 10-4).

One last caution is to be sure that each connector is treated carefully. The voltages that are sent as signals are often small and can be easily altered by dirty or poorly fitting connections. Also, the connector pins of the electronic chips are very fragile. You will not get many chances to put one in place. Be sure that the pins all align correctly with the socket before you try to connect them.

Sensors

Sensors can be grouped according to the output signal they produce. The more common types include switches, variable resistors, induction

Chap. 10 Electronic Service 107

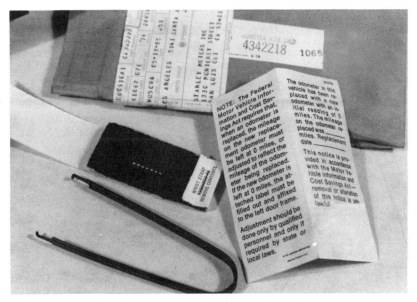

Figure 10-4 Electronic component on static bag with inserting tool.

voltage generators, thermal generators, electrochemical generators, and so on.

Sensors that use coils to produce an induced voltage can be tested by checking the coil leads for the correct resistance and for shorts and grounds. Factory specifications will be needed to determine if the values noted when testing are within the correct range (Fig. 10-5).

Sensors outputting a voltage may be tested to determine if the output is in the correct range. An example would be looking at the output of an oxygen sensor as the engine is running at normal operating temperature. The output voltage should vary from positive to negative

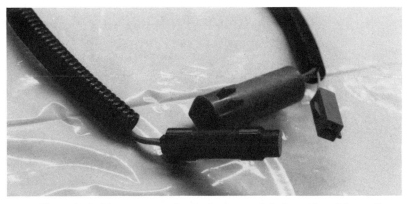

Figure 10-5 Meter connected to leads on a crankshaft-position pickup coil.

and back fairly often if the fuel mixture is near stoichiometric and the sensor is functioning properly. Another example is to determine the output voltage of a permanent-magnet generator by allowing the engine to drive the wheels while they are jacked up off the floor. The output voltage of the generator should correlate with the speedometer and voltages given in the workshop manual (Fig. 10-6).

The basic procedure is first to determine what specifications are available. Then test for those specifications. Read the workshop manual for the correct testing procedure. Follow that procedure exactly until you understand how the system functions. Shortcuts often cause short circuits.

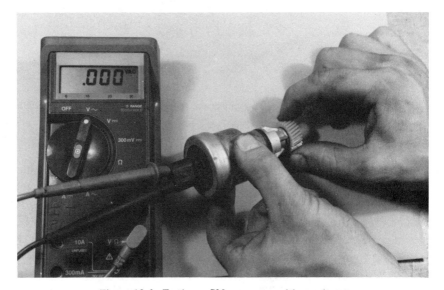

Figure 10-6 Testing a PM generator with a voltmeter.

Another possible testing technique is to introduce the element the sensor is testing for and see if the system reacts to the input. For example, you could be observing the timing of an engine with a timing light and then introduce a knock near the knock sensor. Simply rap on the engine very near the knock sensor while the engine is running. The timing should retard if the sensor and electronic system are functioning properly (Fig. 10-7).

Temperature sensors may often be tested for their resistance with the engine cold and later with the engine warmed up. The resistance should decrease as temperature increases and in both cases be within the factory specifications.

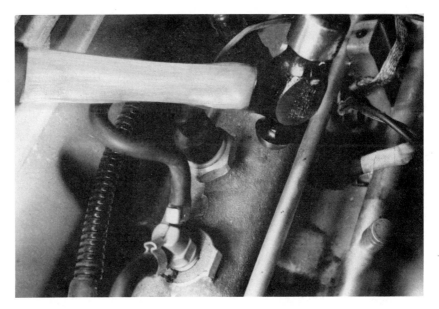

Figure 10-7 Tapping on manifold next to knock sensor with 8-oz hammer.

The questions to be answered are:

1. If an input voltage is required to operate the sensor, is it there and of the correct value?
2. Is the output voltage correct for the operation of the sensor?
3. Does the sensor have the correct resistance?
4. Does the operational system make changes as expected when a specific condition is introduced?
5. Is there a fault code for that specific area?
6. Is the self-diagnosis system operating properly?

When these questions are answered, you will, hopefully, have found the problem. You will need to understand what type of sensor it is, what type of test can safely be performed, and what the expected results might be.

Actuators

Many actuators are either solenoid or electric motor devices. The windings of these devices can be tested for resistance, shorts, opens, and grounds. An accurate ohmmeter is a must for these measurements. You will also need the factory "specs".

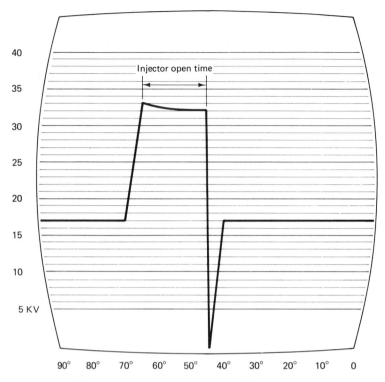

Figure 10-8 Oscilloscope pattern of fuel injector voltage.

Voltages to these devices must be measured accurately in the dynamic or "on" state. Another possible test is to measure the current used by the actuator as a check on the running state of the system.

Another test of an actuator device is the use of an oscilloscope to compare the pattern created by each actuator. An example of this would be to use the primary section of a standard auto type oscilloscope to "look" at the pattern made by each fuel injector on a given engine. Compare each injector oscilloscope pattern for differences. The differences could be a variation in the width, height, or actual pattern shape. These differences may be quite apparent (Fig. 10-8). This test is accomplished by connecting a lead from the ground wire of the injector to the primary pattern pickup of the oscilloscope (Fig. 10-9).

Connectors

The variety of electrical connectors is expanding at a rapid rate. Many of the new connectors use some type of lock to keep the connector together in spite of vibration, pulling on wires, and mechanical work

Figure 10-9 Connecting primary lead of oscilloscope to fuel injector. Courtesy B-O-C General Motors.

done in the area of the connector. The connector must make and keep the electrical current moving without threat of corrosion, looseness, dirt entry, or deterioration over time.

The first thing you have to learn when servicing electronic components is how to disconnect all the various types of connectors without damaging them. Many require prying back a locking or catching tab before pulling the two halves apart. Some connectors use some type of sealant or soft insulating material (Fig 10-10).

Most new connectors can be put together in only one way. Look carefully at the pins of the male connector to ensure that they are clean, straight, and correctly located in their housing. Inspect the female connector to be sure that no foreign material has entered the pin slots. When the connectors are slid together, ensure that the locks are fully engaged (Fig. 10-11).

Figure 10-10 Close-up of electronic lead connector lock.

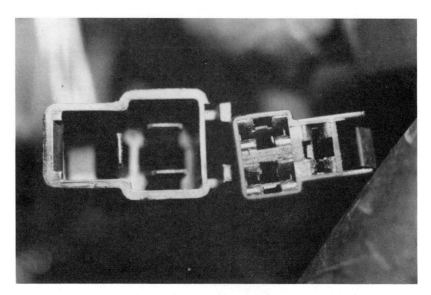

Figure 10-11 Close-up of male and female connectors.

Controls

The major control system of the automobile is the driver. The first rule of good service is to determine if a mechanical or electronic problem exists. Find out from the vehicle operator what symptoms he or she thinks the auto evidences. The problem might be that the operator does not understand how the system is supposed to work. In other words, first make sure that there is a vehicle problem.

Servicing most control systems or input devices consists of testing switches for "on" and "off" resistance, inspecting connectors carefully, and ensuring that there is a voltage input to control. Use your voltmeter to find the value available to the switch. Determine if that same value is available at the output side when the switch is operated. A final service procedure, when other options fail, is to substitute a new part.

Instrumentation

The first service procedure for any instrumentation system is to understand exactly how the instrument operates. Review, if necessary, the service manual's introductory theory section on that specific instrument package.

If you need to handle or work near any instrumentation, take the proper electrostatic-discharge precautions. Do not touch any individual electronic component with your fingers. Ground tools such as screwdrivers and nutdrivers just before you start removing an instrument section. Carefully observe the connection and routing of any ground straps so that they are replaced properly when the unit is reinstalled.

Sliding around on fabric seats or carpet can build up a new static charge. Sheepskin seat covers are particularly good at building static discharge under the right climatic conditions. Do not let that voltage get near any instrument system.

Let the built-in self-diagnostic capability of the unit help you. Get the diagnostic codes and study them before you start working on the car. Have those codes right in the auto with you. Also, take a pad and pencil to write down the codes so that you will not have to memorize them as they are being displayed.

Follow the manufacturer's diagnostic "tree" exactly. The engineers that built, tested, and developed the diagnostic procedures have many reasons for the way they want you to approach diagnosis. Remember the saying "Shortcuts lead to short circuits." Also remember how important it is to approach electronic systems with the right attitude (Fig. 10-12).

Carefully inspect every connection. The currents are very small in most electronic instrumentation packages. These connectors have to be very good. Look for "backed-out" terminals, corrosion, or other foreign material and physical damage to the connectors. Some factory-approved testers are available if you can afford them. Specific inputs are given to the instrument being tested to see if it responds correctly. When inputs to an instrumentation package are good, connectors look good, but if the unit still does not function properly, you will probably have to send it to a factory-authorized repair center.

Be sure to take out the odometer chip and store it properly before you send the instrument cluster someplace for repair or replacement. That chip probably will not be returned with the new package and purchasing a new odometer chip will add to the expense. Also, the new chip will have to be programmed to the correct mileage to be legal. This operation can only be done by the repair center (Fig 10-13).

One last reminder! Approach each electronic service job with the most positive attitude possible. The time spent thinking and being careful will pay off.

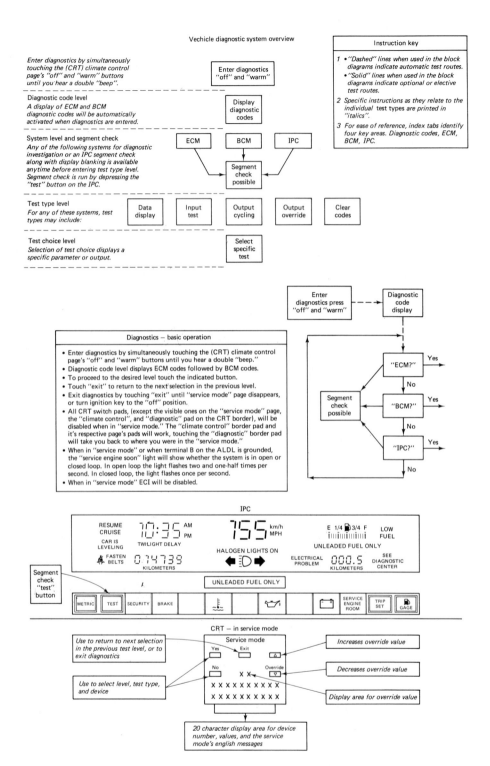

Figure 10-12 Manufacturer's diagnostic "tree." Courtesy B-O-C General Motors.

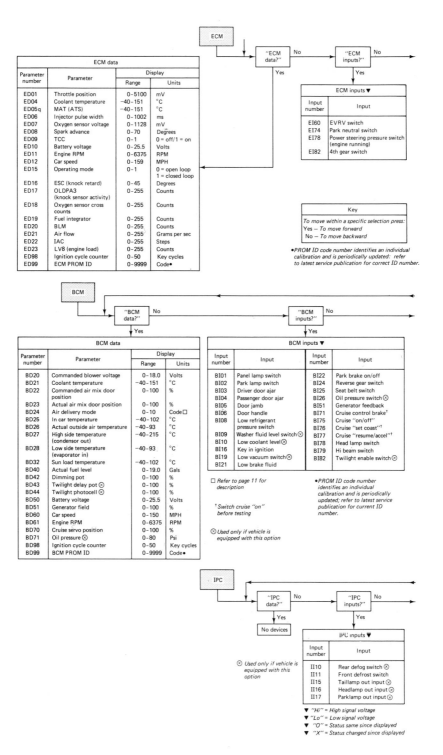

Figure 10-12 (continued)

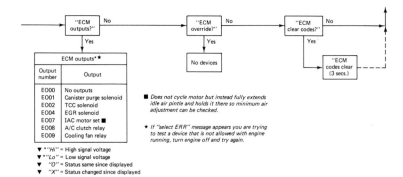

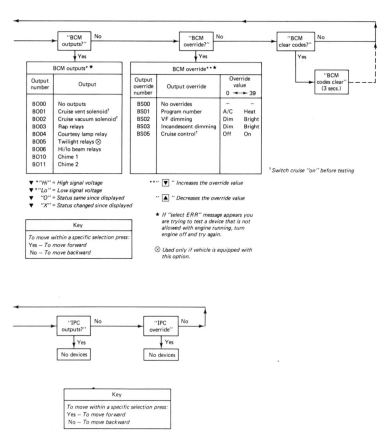

Figure 10-12 (continued)

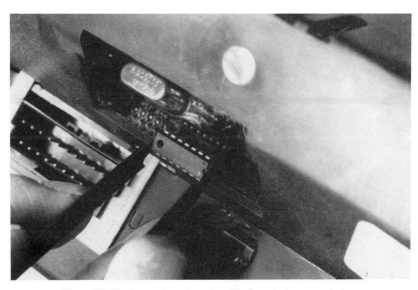

Figure 10-13 Removing odometer chip from instrument cluster.

Glossary

Actuator an electromechanical device that produces motion, e.g., a solenoid.

Ampere also called amp; the electrical unit that describes the number of electrons going past a given point during one second. Equivalent to one coulomb per second. The current produced by one volt applied across one ohm resistance.

Aneroid actuated by air pressure.

Backlighted light placed behind a display and allowed to pass through it and be seen from the front.

Base the control terminal of a transistor. The center section of a transistor.

Buffer A circuit used to change an analog signal into a digital signal.

Collector the output terminal of a transistor.

Computer a set of electronic circuits capable of following a program, accepting specific inputs, processing that data and producing specific outputs.

Computer Clock a basic time input to the control unit of the microprocessor.

Computer Program a series of specific directions the processor follows to direct the other components of the computer.

Conductor a material that allows electrons to move easily through it.

Control Unit also called the CU; is the part of a computer that directs the other parts by following the sequence in the program.

CRT cathode ray tube; like the vacuum fluorescent tube that is the display part of a television set.

Data Line an electrical conductor capable of being connected to many electronic components that need and want to share information.

Dedicated Test Equipment test equipment that can be used on only specific make or model vehicles. Performs limited very specific tests.

Depletion Layer a zone between the bonded P-type and N-type layers where the positive and negative electrical charges are balanced.

Dichrolic a liquid dye mixed with the crystalline liquid in an LCD to produce a colored background.

Diffusion Potential the voltage that exists across the depletion layer in a diode.

Diode a piece of N-type semiconductor material bonded to a piece of P-type semiconductor material. A two component semiconductor having two terminals.

Direct Fire Ignition an ignition system that uses a coil for each one or two cylinders and no distributor.

Doped having added a very small quantity of some other material to the base material.

Electrical Electricity conducted through conventional metallic types of conductors.

Electrochomism the characteristic of being able to change color when a voltage is applied.

Electronic electricity conducted by components other than metallic materials.

Electronically Controlled Braking A system of sensors and actuators that compares vehicle wheel speeds and releases hydraulic pressure to different wheels to maintain the optimum slippage rate.

Electronic "Chips" small thin pieces of pure silicone crystalline material that have been altered to become complex electronic circuitry.

Electronic Memory storage of informational bits in the RAM or ROM or other memory areas of a computer.

Electronic Navigation a system of displaying the vehicle's current position on a CRT display of a map. Various map scales may be chosen by the operator.

Glossary

Electronic Torque Control an electronic system that senses the torque applied to each vehicle wheel and limits it to avoid excessive wheel slippage.

Electrophoresis the theory of movement of charged particles in a liquid when a voltage is applied.

Emitter the power input connector of a transistor.

Epitaxy depositing semiconductor material on a semiconductor base.

Fiber Optics the transmission and receiving of light signals sent along fine glass fibers. Cleaner signals are received as compared to metallic conductors because the electrical induction of additional energy into the glass fiber can not occur.

Forward Bias a negative potential (voltage) connected to an N-type material in a semiconductor component.

Global Bus all components attached to the conductor would receive all the sensor and other signals sent.

Hall Effect an almost square waveform generated by a particular solid state component when fluxed by a magnetic field.

Holographic an image produced by multiple light sources in air rather than on a specific surface.

Induction also called electromagnetic induction; the process of producing electrical energy by passing magnetic lines of force through a conductor.

Insulator a material that electricity has a very difficult time moving through.

Integrated Circuit a complex complete circuit manufactured within one very small continuous piece of semiconductor material. It is possible for more than forty thousand individual components to be manufactured into one IC "chip."

Light Emitting Diodes also LED; a two-part solid state device that emits light when current of the correct polarity is passed through it.

Liquid Crystal a liquid material that exhibits the organized molecular structure usually associated with solid crystals.

Liquid Crystal Display also LCD; using electricity to change the normal crystalline structure of that specific liquid to allow light to pass through or be blocked by it. A display system that reflects light in a specific format to convey information.

Main Bus System a standardized plug and conductor system used to interconnect many components. Many plug-in slots would be located in one central easily accessible place.

MAP manifold abosolute pressure.

Matrix an orderly arrangement of parts or components.

Microcircuitry using photographic reduction to produce extremely small electronic circuitry.

Mircrocomputer a microprocessor with memory and input/output circuits added.

Microprocessor microcircuitry that contains both the computer control unit and the processor.

Minicomputer a microcomputer with a power supply, software, keyboard and CRT added.

Multiplexing sending multiple signals over the same conductor at the same time. This can be accomplished either by very rapid sequencing or by using different frequencies.

Nematic Liquid Crystal a specific alignment of the crystalline structure in a liquid crystal. The nematic structure is the one used for electronic display systems.

N-type Semiconductor an insulator with four electrons in the outer electron shell doped with a material having only three electrons in the outer shell.

Ohm the electrical unit of resistance to current flowing through a conductor. Similar to friction in the fluid and circuit in a hydraulic system.

Ohm's Law states the relationship between voltage, current and resistance. Voltage equals the current times the resistance.

Out-of-Parameter not within the specifications expected to be received.

Parallel Circuit a circuit where the current may be divided among two or more loads.

Passive Display a display system that does not produce light.

PGD planer gas discharge display.

Plasma Emission a gas glowing or emitting light as its molecules release energy.

Polarized Light light that has a particular waveform orientation.

Processor the part of the computer that accomplishes math functions and causes branch functions to occur.

PROM programmable read only memory.

Glossary

P-type Semiconductor an insulator with four electrons in the outer electron shell doped with a material having five electrons in the outer shell.

Quiet Ground an electrical ground on a vehicle that is free from voltages produced by other components that are part of the vehicle. Stray tiny voltages may be read as sensor inputs and produce unwanted computer outputs.

RAM random access memory.

Rectification changing alternating current (AC) to direct current (DC).

Resistance Sensing an electrical or electronic circuit that measures the electrical resistance of a part, e.g., the ignition key, and determines if that resistance is correct so that the rest of the circuit may become functional. Used as a theft deterrent.

Resistor a material that does not allow electrons to travel easily through it.

Reverse Bias a positive potential (voltage) connected to an N-type material in a semiconductor component.

ROM read only memory.

Semiconductor a good insulator that has had a small amount of impurity added to allow electrical conduction.

Sensor an electronic, electric or electromechanical device that correlates a physical phenomenon to an electrical output.

Series Circuit an electrical circuit where each load or component is linked one after the other. The same current is passing through all the loads in the circuit.

Solenoid the combination of an electromagnet, a spring and an iron armature. When the electromagnet is activated, the armature moves toward it. The spring returns the armature when the electricity is off.

Speech Synthesizing electronic circuitry coupled to a speaker to produce sound that is recognizable as voice patterns or words.

Threshold the amount of electrical energy it takes for a display element to start emitting light or otherwise become activated.

Touch Sensitive fine electrical wires placed in a grid format near the surface of a object. The electrical resistance of a specific part of the grid is changed by the touch of the human finger. That resistance change is used as an electronic input. The grid could be placed on the display end of a CRT.

Transistor a semiconductor device consisting of three bonded semiconductor regions. The regions may be arranged as either PNP or NPN.

Transmissive Display a display that allows light to come through from the back to the front.

Vacuum Fluorescence light emitted from a specific material which is being bombarded by electrons that are accelerated through a vacuum.

VF vacuum fluorescent display.

Volt the electrical unit for force; similar to pressure in a hydraulic system.

VOM volt ohm meter.

Voltage the force that attempts to move electrons through a conductor.

Wiegand effect the particular properties of a special magnetic wire that produce an almost square waveform in the generated voltage as a magnetic field passes through it.

Index

A

Actuators, 4
  function, 96
  testing, 109–10
  types, 96
Addresses, 55-56
Air mass, 65
  measure, 65
Alpha/numeric display, 33, 34, 45
Ambient light, 30, 34
Ampere, 13
Analog, 42
Aneroid linear variable differential transformer, 80
Anode, 43, 44, 45

B

Backlighted, 73
Barometric pressure, 63
Base, 24
BCM, 61
Binary code, 56
Bit, 56
Body management, 71-72
Buffer, 53, 89, 94-95
Byte, 56

C

CU (*see also* control unit), 51
Canister purge, 68
Capacitor-capsule MAP, 81
Capacity-type humidity sensor, 88-88-89
Cathode, 43, 44
Cathode ray tube:
  analog control, 47
  theory, 46-49
Charcoal Canister, 68-69
Chip, 26, 27
  display, 3
Circuit, 14, 25
  parallel, 15
  series, 14
Collector, 24
Communication and display, 7

Computer, 50-57
  addresses, 55
  basics, 50
  block diagram, 51
  clock, 50
  firmware, 57, 59
  input/output, 52-54
  memory, 3, 51, 52, 56-57
  operation, 55
  processor, 52
  program, 55, 57-59
  registers, 55
  software, 57
  volatile memory, 57
Conduction, 20
Conductor, 16-17
Connector service, 110-11
Cranking motors, 6
Crankshaft position sensor, 78-80
CRT (*see also* Cathode ray tube), 6, 8, 46, 57, 73
Current, 13

D

Data line, 3-4
Depletion layer, 22, 24
Detonation, 63
  sensor, 63
Diagnostics, 8
Diagnostic Tree, 114-17
Dichrolic LCD, 37
Diffusion potential, 22
Digital, 42
Diode, 21
  light emitting, 39
  operation, 22
  photosensitive, 87-88
  rectification, 22
Display system, 42
Doped, 16, 20

E

ECM, 61
EGR, 67-68
Electrical, definition, 1
Electricity, 11-15, 17
  basic theory, 12
Electrochromic display, 37
Electromagnet, 11
Electromagnetic induction, 11
Electromotive force, 13
Electron, 11, 12
  movement in a material, 13
  shell, 18-20
Electronic:
  circuitry wired into composite
    circuitry, 2
  definition, 1
  engine control, 6
  fuel injection, direct, 6
  ignition, direct fire, 7
  newer technologies, 3
  sensing, 6
  service, 102
  steering control, 5
  torque control, 5
  theory, 16
Electrophoretic display, 37-38
Electrostatic discharge, 113
Emitter, 24
Engine control, electronic, 6
Engine management, 60, 62, 70
Engine sensors, 76
Epitaxy, 26, 28, 29
Exhaust gas recirculation, 67-68

F

Fault code, 70
Fiber optics, 7
Forward bias, 23

Frequency, 7
Fuel injectors, 65
Fuel level sensor, 92

G

Gas discharge display, 43-46
Global bus, 7-8

H

Hall effect, 79, 91
Hole theory, 20
Holographic, 8
Hot-wire airflow sensor, 85-86
Humitidy sensor, 88

I

IC chip, 26, 39, 55, 74
Idle speed control, 69
Ignition, direct fire, 7
Induction, 11
Instrumentation displays, 72-73
Insulator, 13, 17, 19
Integrated circuit, 25-29
  computer use, 54-55
  construction, 26-29
  mask, 27
Ionization, 43

K

Keyboard input, 4
Knock sensor, 83

L

Light emitting diode, 39-40, 73
Light emitting displays, 39
Light-modulating display, 30
Liquid crystal display:
  backlighted, 36, 73
  clock, 30
  construction, 32
  dichrolic, 37
  illustration of, 3
  polarized, 35
  power required, 35
  theory, 31-33
LCD (*see also* Liquid crystal display), 8, 35, 44
LED (*see also* Light emitting diode), 40-41, 44, 87, 90-91

M

Magnetostrictive knock sensor, 83
Main bus, 7
Manifold absolute pressure sensor, 80-81
Mask, 27
Matrix, 20, 40-42, 45
Microcircuitry, 2
Microcomputer, 55
Microprocessor, 3, 4, 55
  body system, 4
  climate control, 4
  cruise control, 4, 72
  ignition system, 4
  uses, 60
Minicomputer, 55
Multiplexing, 7, 40-43, 45

N

Navigation, 8, 48, 73
Negative image, 36

Nematic liquid crystal, 31
Neon, 43
N-type semiconductor, 20

O

Ohm, 14
Ohm's law, 14
Oxygen sensor, 65, 77-78

P

Passive display, 30
PGD (*see also* Planer gas discharge), 44
Phosphor coated, 44, 45, 47
Photodiode, 87
Planer gas discharge, 44
Plasma display, 43-46
Polarized construction, 35
Programmable read only memory (PROM), 56
P-type semiconductor, 20
Purge solenoid, 69

R

Random access memory (RAM), 57
Read only memory (ROM), 56
Read/write Memory (R/W), 56
Rectification, 23-24
Registers, 55-56
Relay, 96-97
Resistance, 14
Reverse bias, 23
Ride control, 6

S

Scanning, 47
Segmented numerical display, 2
Self diagnostics, 104
   construction, 33
Semiconductor, 16, 18-21
   N-type, 20
   P-type, 20, 24
   salt-doped, 88-89
   temperature sensitive, 84
Sensors, 61
   absolute pressure, 63
   airflow, 65, 85-87
   barometric pressure, 64
   buffers, 94
   coolant temperature, 64
   crankshaft position, 62, 65, 78-80
   crankshaft speed, 70
   detonation, 63
   fuel flow, 87-88
   fuel level, 92-94
   humidity, 72, 88
   knock, 83
   manifold absolute pressure, 80-81
   manifold pressure, 62, 80
   oxygen, 65-66, 77-78
   photo, 72
   steering position, 90
   suspension height, 72
   temperature, 72, 83-85
   testing, 106-9
   throttle position, 62, 76
   vehicle speed, 70, 89-90
   wheel speed, 72
Sensor signals, 4
Serial data line, 5
Service, 102
Service philosophy, 105
Silicone chip, 27
Sine wave, 23
Software, 57
Solenoid, 97-98
Solenoid valves, 98-99
Sound-alert, 74
Speech synthesizing, 4
Steering control, electronic, 5
Steering position sensor, 90-92
Stepper motors, 99-100

Index

Strain gauge pressure sensor, 81
Suspension height sensor, 89-91
Switch:
  input, 4, 72
  signals, 92
  throttle position, 62

T

Telltale panel, 73
Theft deterrent, 75
Thermal actuator, 100-101
Thick film deposit, 92-94
Threshold voltage, 41-42
Throttle position sensor, 76
Torque control, electronic, 5
Torque motor, 97-99
Touch control, 3, 49
Transistor, 24-25
Transmissive display, 36

U

Ultrasonic sensor, 87

V

Vacuum fluorescent display, 46, 73
  clock, 2
  theory, 44-45
Variable inductor diaphragm sensor, 82
Vehicle speed sensor, 70, 89-90
VF (*see also* Vacuum fluorescent), 44, 46, 73
Voice-alert, 74
Voltage, 13
VOM, 102-4

W

Wheel speed sensor, 88
Wiegand effect, 79
Wire-would fuel level sensor, 93

Y

Yoke, 47